Writing a Science PhD

Research Skills

Authoring a PhD
The Foundations of Research (3rd edn)
Getting to Grips with Doctoral Research
Getting Published
The Good Supervisor (2nd edn)
The Lean PhD
PhD by Published Work
The PhD Viva
The PhD Writing Handbook
Planning Your Postgraduate Research
The Postgraduate Research Handbook (2nd edn)
The Professional Doctorate
Structuring Your Research Thesis
Writing a Science PhD

Teaching and Learning
Series Editor: Sally Brown

Access to Higher Education
Coaching and Mentoring in Higher Education
Facilitating Work-Based Learning
Facilitating Workshops
For the Love of Learning
Fostering Self-Efficacy in Higher Education Students
Internationalization and Diversity in Higher Education
Leading Dynamic Seminars
Learning, Teaching and Assessment in Higher Education
Learning with the Labyrinth
Live Online Learning
Masters Level Teaching, Learning and Assessment
Reimagining Spaces for Learning in Higher Education
Successful University Teaching in Times of Diversity

Writing a Science PhD

Jennifer Boyle and Scott Ramsay

© Jennifer Boyle and Scott Ramsay, under exclusive licence to
Springer Nature Limited 2019

All rights reserved. No reproduction, copy or transmission of this publication may be made without written permission.

No portion of this publication may be reproduced, copied or transmitted save with written permission or in accordance with the provisions of the Copyright, Designs and Patents Act 1988, or under the terms of any licence permitting limited copying issued by the Copyright Licensing Agency, Saffron House, 6–10 Kirby Street, London EC1N 8TS.

Any person who does any unauthorized act in relation to this publication may be liable to criminal prosecution and civil claims for damages.

The authors have asserted their rights to be identified as the authors of this work in accordance with the Copyright, Designs and Patents Act 1988.

First published 2019 by
RED GLOBE PRESS

Red Globe Press in the UK is an imprint of Springer Nature Limited, registered in England, company number 785998, of 4 Crinan Street, London N1 9XW.

Red Globe Press® is a registered trademark in the United States, the United Kingdom, Europe and other countries.

ISBN 978–1–352–00630–8 paperback

This book is printed on paper suitable for recycling and made from fully managed and sustained forest sources. Logging, pulping and manufacturing processes are expected to conform to the environmental regulations of the country of origin.

A catalogue record for this book is available from the British Library.

A catalog record for this book is available from the Library of Congress.

Contents

Acknowledgements x

Introduction xi

1   What the PhD Is, and What It Isn't   1
    Expectations   1
    Some practical realities of PhD study   3
        Uncertainty   4
        Guidance   6
        Expectations of independence   6
        A new relationship to broader, 'real-world' research   6
    What the PhD isn't   7
    Your thesis   8

2   Establishing a Writing Practice   10
    Managing your time   11
        Audit your time   11
        Creating extra time   13
        Are you writing, or are you distracted?   14
        Are you writing, or are you editing?   15
        Are you writing, or are you reading?   15
    Writing environments   16
    Motivation   19
    Keeping a thesis journal   20
    Process and product writing   21
        The process draft   22
        How to benefit from the process draft   23
        The product draft   25
        Benefits of the process and product approach   27

3   Refining and Articulating Your Research Question   29
    The (un)importance of starting with a defined project plan   29
        Research questions vs. hypotheses   30
    Which one should you choose when you write your thesis?   35
    Why is all of this important?   35
        How do they relate to each other?   36
    Don't expect that your project's goals will stay unchanged   36

4	**Finding Literature**	38
	Finding a good, authoritative, wide-reaching (yet subject-focused) database	39
	How does an academic database work?	40
	How do you choose the best database(s)?	40
	Indexing coverage	41
	Functionality	42
	Constructing a rigorous search	43
	Boolean operators	44
	Why would you need to experiment with the best word to use?	44
	Combining separate terms	46
	Bonus search tools	47
	How many results should you aim to get?	48
	Documenting your search strategy	49
	Exclusion criteria	50
	A helpful example of a specialised principle	50
	PICO (Problem/Intervention/Comparison/Outcome)	50
	Searching beyond the literature review	52
	Automated search alerts	53
	MeSH headings (medicine and biomedical sciences)	53
	Keeping abreast of new literature	54
5	**Organising and Keeping Track of the Literature**	56
	General principles of collecting reference information	57
	Reference management software	59
	Good practice when using a reference manager	59
	Getting information about your sources into your reference manager	60
	Checking your reference manager's library is complete and accurate	62
	Getting references back *out*	64
	Creating a reference list	64
	Keeping the list in order	64
	Making reference lists from only some entries in your database	64
	Reformatting references in a different style	65
	Inserting citations into your thesis as you write	65
	Annotating and grouping your references	65
6	**Reading and Critiquing the Literature**	67
	Reading strategies	68
	Choosing which type of articles to read first	68
	Becoming confident navigating articles	68
	Choosing which section of an article to read first	69
	What is critical analysis?	71
	Why is critical analysis important?	72
	How can a student like you be qualified to critique an expert's published work?	74
	On what basis can you critique someone's work?	75
	Sample size and replication	76

	Study design's relationship to the research question	78
	Use of controls	79
	Use of 'optimisations'	80
	Recency of cited material	81
	Authority of the author	82
	Pressures to publish bad science/why bad science gets published	82
	Further questions to ask	83
	Ambiguity and the realities of scientific research	84
7	Structuring Your Chapters	85
	Principles of a good chapter	85
	Typical chapter structure	87
	Variations in chapter structure in the PhD	88
	Introductions	88
	Body chapters	91
8	Writing About the Literature	94
	How many sources do I need?	94
	How often should I use sources?	95
	Which sections of my thesis should include references?	95
	What are the different reasons for citing others' work?	97
	To give credit	97
	To help your reader locate the original source	100
	Plagiarism	101
	How to use the works of others without breaking any rules	102
	Citing articles within articles	103
	Self-plagiarism	104
	Referencing styles	105
9	Structuring Your Sentences	106
	Sentence structure types	106
	Simple sentences	107
	Compound sentences	107
	Complex sentences	109
	Compound/complex sentences	110
	How to use sentence structure to your advantage	111
	Punctuation rules and common errors	115
	Commas	115
	Common comma errors	119
	The comma splice	119
	Semi-colons	121
	Colons	122
	Summary of sentence structure and punctuation	124
	Sentence structure	124

10	**Paragraphs**	125
	Types of paragraph	126
	Common paragraph problems	129
	The topic sentence is too broad	129
	The topic sentence is missing	130
	Lack of interpretation	131
	The paragraph contains too much material	132
	The paragraph doesn't have enough material	133
	Connectives	133
11	**Editing and Proofreading**	136
	The difference between editing and proofreading	137
	How to edit effectively	137
	How to proofread effectively	140
	Common problems	140
	Tips and techniques	140
	General tips for editing and proofreading	142
12	**Making the Document Look Like a Thesis**	144
	Making life easier for your reader	145
	Cross-referencing	145
	Navigating the document	146
	Standard document sections	146
	Placing figures and tables	147
	This all has implications for the process of writing your thesis	148
	Clarity in tables and figures	148
	Panelling	149
	Relating sections to each other	151
13	**Writing for Publication**	152
	Why write for publication?	152
	Things to consider	153
	What makes a good journal article?	155
	Beginning the process	155
	How does peer review work?	158
	Types of peer review	160
	After peer review	161
	Terminology around publication	167
	Other ways of writing beyond the PhD	168
14	**Common Roadblocks During the PhD**	169
	Dealing with your supervisor	170
	Mismatched expectations	170
	Good practice before supervisory appointments	172
	General advice	174

Isolation	174
Productivity	175
Procrastination and perfectionism	177
Procrastination	178
Language issues	182
Life outside the PhD	184
Conclusion	185
Appendix: List of Useful Software and Apps	186
Bibliography	188
Index	190

Acknowledgements

We would like to thank the former PhD students who contributed their recommendations to you, which you'll find spread throughout this book: Mai-Britt Jensen; Fabian Kellermeier; Christine Merrick; Lydia Soraya Murray; Anne Tierney; and Angela Wilson. We also wish to thank the editors at *Nature Research*, who contributed their experiences of what they thought their PhD would be like, and what they actually turned out to be: Dr João H. Duarte, Senior Editor, *Nature Biomedical Engineering*; Iulia Georgescu, Chief Editor, *Nature Reviews Physics*; Marios Karouzos, Senior Editor, *Nature Astronomy*; and Bart Verberck, Regional Executive Editor, *Nature Research*. We would also like to thank our current PhD students. Particular thanks go to Sejal Modha for kindly sharing her abstract writing technique. Finally, thanks to our own PhD supervisors, Dr Peter Dominy and Dr Marilyn Dunn, for helping us shape our own writing practice and style over the course of our PhDs.

Introduction

This book aims to give you the knowledge and skills you need to write your PhD thesis. Undertaking a PhD is a huge task, which involves not only working at the highest academic level, but also transitioning from the role of student to researcher and becoming fully integrated into the academic community. While there are many books available on writing the thesis, none talk specifically to science students, who deal with a range of challenges specific to their discipline.

In this book, we'll talk about:

- how to formulate your research question
- how to gather relevant literature
- how to record literature, and how to critique what you find
- how to discuss existing work in relation to your own
- how to factor writing time into your schedule, and how to make that time as efficient as possible
- how to structure your work, from the smallest scale to the largest, to ensure clear and concise communication
- how publication can be part of your PhD
- how to tackle common roadblocks.

We hope that you'll find the information here practical, easily accessible and relevant to your specific needs. We work with PhD students at every stage of their research, and so the advice we offer is grounded on the needs of those students and the questions they most commonly ask us.

This book assumes that you are writing a 'traditional' thesis – that is, not a thesis by publication. The thesis by publication route is still in the minority in the UK (it's unusual at our own research-intensive institution, for example), but is gradually increasing in visibility as a viable option.

The term 'PhD by publication' can mean two things:

> The first type of 'PhD by publication' refers specifically to the process by which a candidate can submit a portfolio of previously published

work to their institution in order to be considered for the award of PhD. This is usually open only to members of staff or graduates of that specific institution who have built up a suitable research portfolio throughout the course of their career.

The second type, and the type that is likely more applicable to you, is also known as an 'alternative format' thesis, or a 'journal format' thesis. In this kind of thesis, chapters and sections are explicitly written as journal articles, and have usually already been published, or are in the process of publication, or have been submitted to a journal for consideration.

It should be noted that this type of PhD thesis must still be coherent – the chapters/articles cannot be wholly disparate. The candidate is expected to produce an introduction which contextualises the work and makes the connection between the chapters/papers clear. Equally, there needs to be a strong conclusion/discussion/summary which draws the work together. Remember, too, that there is likely to be some additional material required in terms of overall analysis and reflection, and possibly further/more in-depth reference made to related work.

While we cover material that will still be useful to you in terms of writing practice, editing, etc., you should make sure you are familiar with your relevant institutional guidelines and assessment criteria. The format and expected number of papers can vary from institution to institution, and across subject areas within institutions, so you need to be fully informed and completely clear on what is expected of you.

There are many disciplines and specialisms within the sciences. We've tried to keep this book as open as possible in terms of the advice offered and used examples which should, hopefully, be accessible to all. When subject-specific differences might arise, we will flag this in the text to allow you to check your own specific conventions with your supervisor and departmental guidelines.

We also have observations throughout from successful PhD students from across the sciences, reflecting on their experiences of writing the thesis, and offering advice that they wish they had known when they were students. You'll also find insights from several senior editors of scientific journals.

We've tried to make sure that the content here is all guidance that you can immediately put into practice, no matter what stage you are at in your PhD, and – as such – we've kept discussion of the more theoretical issues around some of these topics to a minimum. If you do want to investigate any of these areas in more depth, the bibliography at the end of the book has guidance on further reading that you could undertake on specific topics.

We hope that this book will help you avoid, as much as possible, some of the low points and stresses that can come in the writing of the thesis. Undertaking a PhD gives you the opportunity to spend concentrated time studying something you are genuinely passionate about. We want to ensure that communicating that work to your supervisors, examiners and a wider audience is a task you feel both proficient in and enjoy.

CHAPTER 1 # What the PhD Is, and What It Isn't

Many students we talk to tell us that they spent a lot of their first year confused about what they should have been doing, or wondering whether what they were doing was 'right'. Many felt other students simply knew more than they did about everything from internal university procedures, to writing the literature review, to managing supervisory relationships. We'll try to give you an idea of what these are each likely to require of you.

We'll discuss what students tend to assume, and how this aligns with what institutions actually want. We'll examine topics in the context of how they'll introduce you to, and acculturate you in, the wider academic community. This chapter will also ask you to consider how 'embedded' you feel in your department, and get you thinking about what you might do in order to feel more of a part of that wider research community.

We'll also look more specifically at the thesis itself and explain how it differs from undergraduate work in terms of depth, scale and content. We'll also try to make sure that you understand the importance of relating your work to the wider literature base, and of reflecting on the novel contribution you'll make.

Expectations

The PhD is the highest level of degree that you can undertake, and almost certainly the most demanding piece of work that you'll have taken on in your life so far. It's an independent research project, and, as such, it's the means by which you demonstrate your ability to manage a research project and enter the academic community.

The Quality Assurance Agency for Higher Education (the QAA) is the body which maintains standards within Higher Education in the UK. Their qualifications framework formally describes the PhD as follows:

> Doctoral degrees are awarded to students who have demonstrated:
>
> - the creation and interpretation of new knowledge, through original research or other advanced scholarship, of a quality to satisfy peer-review, extend the forefront of the discipline and merit publication
> - a systematic acquisition and understanding of a substantial body of knowledge which is at the forefront of an academic discipline or area of professional practice
> - the general ability to conceptualise, design and implement a project for the generation of new knowledge, applications or understanding at the forefront of the discipline, and to adjust the project design in the light of unforeseen problems
> - a detailed understanding of applicable techniques for research and advanced academic enquiry.

> Typically, holders of the qualification will be able to:
>
> - make informed judgements on complex issues in specialist fields, often in the absence of complete data, and be able to communicate their ideas and conclusions clearly and effectively to specialist and non-specialist audiences
> - continue to undertake pure and/or applied research and development at an advanced level, contributing substantially to the development of new techniques, ideas or approaches.

> And holders will have:
>
> - the qualities and transferable skills necessary for employment requiring the exercise of personal responsibility and largely autonomous initiative in complex and unpredictable situations, in professional or equivalent environments.

If you search for postgraduate codes of practice and associated regulations on any university website in the UK, you will usually find these guidelines in slightly varying wording but with the same underlying principles. You are developing the ability to act as an autonomous researcher, with the guidance and support of your supervisory team.

Some practical realities of PhD study

Insights from a journal editor

PhDs are where true science happens, or so I thought. Before I began my PhD, I had illusions of grandeur, of scientific breakthroughs and – I kid you not – Nobel prizes. The general topic of my research was black holes. It doesn't get grander than this! And yet, the first task I was assigned was to read a bunch of papers – literature research. The next steps included wading through hundreds of references, collating information from different sources and eventually building a database – not very grand, not very ground-breaking. It was instead often boring, usually tiring, at times even defeating. In retrospect, it was also important: a basis without which any further progress could not have been accomplished.

The day came when I was called to make my first presentation about my PhD research. This was my chance to shine, I thought. My slides were impeccably designed, dotted with big question after big question, fundamental unknowns about the Universe, scientific conundrums…and about black holes. I was already at slide 25 (of a 30-minute presentation) before I 'stooped down' to explaining the literature search I had done, the information I had collated and the database I had built. Needless to say this particular version of my presentation was never seen by anyone other than by my PhD supervisor. After slide 5 or so, she had already concluded that it was all wrong and I had to redo it, I had to sharpen it, I had to focus the slides on *my* work, *my* research.

This clash between my expectations of what doing a PhD would entail and what my PhD was actually about followed me well into the second year of my three-year graduate programme. It was a dissonance that made me feel at the same time inadequate – am I failing to do ground-breaking science? – and unfulfilled – when will the menial tasks give way to true science? And while I cannot preclude a major scientific breakthrough happening in your PhD, I can certainly tell you that usually a PhD is nothing more than an appetizer, a training course meant to teach you how to do (independent) research, a bite-sized – your bite size mileage may vary! – problem to sharpen your scientific skills on.

One should not regard doing a PhD as inventing the wheel, but perhaps more like measuring the curvature of a wheel. Or rather, the curvature of a very specific segment of a wheel. One wheel of a many-wheeled vehicle. And that's okay!

Dr Marios Karouzos (Senior Editor, *Nature Astronomy*)

Uncertainty

The PhD is not supposed to be the *perfect* research project. It is *a* research project. You'll develop the abilities and the skills we've described above over the course of the whole PhD. As such, you should recognise that it is a learning experience. You're not expected to demonstrate proficiency from day one, so don't be afraid to acknowledge when you're unsure of things.

You'll encounter gaps in your knowledge, seemingly insurmountable problems, unexpected results, and roadblocks that force you to completely rethink your approach. The QAA guidelines above refer to these as 'complex and unpredictable situations'. This does not mean that your research is going badly. This just means your research is like the majority of research being carried out everywhere else.

Look at this real-life example of a 'complex and unpredictable situation'.

Insights from a journal editor

I started my PhD studies with a research project on the adaptive immune system of zebrafish. These small fish had been studied for decades in the field of developmental biology, as their high-throughput potential and transparency made it possible to do large functional studies, but they were mostly 'uncharted land' in terms of our knowledge of their immune system.

My supervisor and I set to explore the main organs of interest from an immunology perspective and attempted to establish disease models in the fish that would have relevance for higher organisms. However, after several months of roadblocks and unexciting results, we decided to scrap the project and set on a different research direction, changing to a more conventional animal model in the field and to more cutting-edge questions.

This was a radical decision. At first I was shocked and very worried: we lost several months' worth of time and effort apparently for naught, and it wasn't clear to me if changing projects would set us in a more productive path. But by the end of my studies I came to appreciate the importance of that decision. The change meant that I acquired a broader set of skills and experience, and put us on a research track that followed fresh discoveries in the field and made much better use of the local technical expertise. The new project was fruitful and led to a publication.

Dr João H. Duarte, (Senior Editor, *Nature Biomedical Engineering*)

The example above not only gives a good sense of the type of situation you might encounter, and how these difficult situations are actually the ones that help you develop most as a researcher, but it also highlights the importance of a good relationship with your supervisors.

Guidance

Your main guides throughout your PhD will be your supervisors. While many institutions are likely to have guidelines on roles and responsibilities in the supervisory relationship, there can be large variation within that relationship. To take an example, 'regular meetings' might mean very different things to different people: a daily 20-minute conversation, a weekly meeting, a formal monthly meeting or a two-monthly Skype meeting, depending on your respective commitments. Some supervisors will see it as their role to offer career advice from the outset, and others might expect you to ask them if you want guidance in that area. Some supervisors might want to give you detailed feedback on every aspect of your writing, and others simply might not see this as part of their remit. We'll think about how to negotiate these differing expectations in Chapter 14.

Expectations of independence

No matter what level of support you have from your supervisor(s), a PhD will involve more *independent* work than you're used to from your undergraduate studies. The guidelines we quoted above repeatedly highlight the importance of independent critical skills and personal responsibility. While your supervisors might provide guidance when it comes to carrying out the literature review, for example, and while your institution might offer additional training in how it should be produced, you're still expected to be able to critically evaluate what you find, and to be able to offer ideas on how your research project fits in to the wider field, how it relates to existing work and the impact it is likely to have.

If you come from a different academic culture, you might find that this new set of expectations takes some additional adjustment. Equally, it's important to remember that your supervisor knows and expects that your relationship is a teaching relationship – although you will, by the end of your studies, communicate on a more level playing field. It's important that you ask your supervisor questions when you need to.

A new relationship to broader, 'real-world' research

You might also have to get to grips with thinking of your work in a new light: the impact it will have on the wider academic community.

Publication is increasingly a reality for PhD students throughout their studies, and thinking about how you communicate with a variety of different audiences (particularly if you undertake interdisciplinary work, for example) is likely to take some adjustment.

It is common for students to feel quite lost at the beginning of the process. This is a typical comment:

Insights from a researcher

I arrived a little later than I had expected and then had to spend some time sorting out accommodation, so I missed a couple of induction talks. I've had mostly email contact with my supervisor, until we met for the first time last week. They've told me about the first year requirement, which is writing an overview of the relevant literature – and encouraged me to get started on that. They've said they'll look at it once I have a decent first draft – but I'm not really sure what a 'decent' first draft would look like, or how much detail they're expecting of me.

What the PhD isn't

A PhD isn't a research process where you can rightfully *expect* to arrive at the answers. As we mentioned from the national guidelines we outlined above, you'll be working at the forefront of human knowledge and understanding. Things won't go smoothly, and your supervisor won't know the answer to all of the questions or solutions to all of the problems that you have. If they did, there probably wouldn't be much of a need to be doing the investigations.

The PhD also isn't a guaranteed track to a certain number of publications. No one can predict how well your research will go, and if you're working with a very highly integrated research group then it may be the case that your research contributes a small piece of a much larger puzzle.

In such cases, you might find that your experiments go well enough only to produce a figure suitable for publication right at the end of your time in that lab, and that a good paper suitable for publication in a journal with a high impact factor depends on the inclusion of a colleague's work. If this happens, you might find yourself waiting until after you've completed your PhD and moved on before your supervisor feels that the whole lab team are ready to submit their work to a journal. On top of that, there's no guarantee that the paper will be accepted. If it is, you'll undoubtedly be asked to make a number of revisions as a result of peer review, and that might involve more lab work. If you've moved to another lab, you might be waiting for someone else to repeat and refine the experiments you had originally carried out. In short: publication can be a long, complicated, slow-moving process, and signing up for a PhD is not necessarily the same as signing up for one or more entries on your publication record.

Your thesis

There's a high likelihood that your thesis will be the longest single piece of writing you ever produce. It's not uncommon for a PhD programme to involve three to three-and-a-half years of research time followed by six months of writing-up time. Have you ever spent half a year working on a single file? An average thesis might be between 150 and 200 pages (A4 sized, with lines at 1.5× or 2× the standard spacing, and printed on one side of each page only) – or, if you prefer to think in word counts, 70,000–100,000 words. How many words can you write each day? How many pages? A standard A4 page with *single* spacing can hold about 500 words. If you set yourself a goal of 1,000 words per day (i.e. around two pages of single-spaced text on A4 paper with conventional, non-PhD-thesis formatting), it'd take you 100 days just to write your first draft. You're given a long time because it's a large undertaking.

However, the reason it's so difficult isn't just size. It's also to do with the high degree of both internal complexity, where you cross-reference between things like results and methods sections, and external complexity, where you make connections with what's known outside of your own experimental and theoretical work.

The word 'thesis' essentially means the same as 'argument'. It's possible to think of it as an extended report of what you've found, but if that's the picture you have in your mind then you're setting yourself up to have a more difficult job later. Instead, try to think primarily about writing something that adds new understanding to a large and complex situation. That size and complexity comes from the literature that already exists, and your job is to show how your results fit within that landscape. If you focus on doing a very good job of reporting your findings but neglect to tie the importance of those to the wider field, your document becomes less of a 'thesis' and more of a simple report.

The type of writing expected of you in your PhD should not be completely unfamiliar territory, though: the same essential rules apply from the types of report that you likely produced in your undergraduate studies. (For a more fundamental primer on these general scientific writing concepts, you may find our previous book useful: *Writing for Science Students*. Boyle and Ramsay, 2017) However, managing to maintain coherence and structure over a much longer word count can be challenging. Further, while you are likely to have deadlines imposed by your institution – such as the annual progress report, for example – you are likely to be managing the whole writing process alone over a much longer period of time. As such, it is important to develop and sustain regular working habits to make sure you make good progress.

CHAPTER 2 Establishing a Writing Practice

In our experience, science students rarely think of themselves as writers. However, the sheer length of a PhD thesis, averaging between 50,000 and 80,000 words, means that you are likely to write more than the vast majority of the population. This means that in the eyes of other people, you are a writer. If you decide to continue in academia, then you'll write even more: journal articles, proposals, grant applications, etc. You will be expected to write well, and to write frequently.

Whatever attitudes and ideas on writing you may have brought with you from undergraduate and master's studies, the thesis gives you an opportunity to build an entirely new writing practice that will stand you in good stead throughout the PhD and beyond.

Establishing a writing practice is about making writing a habit, discovering what routine works best to maximise your productivity, and embedding this routine in your schedule. This chapter will guide you through developing a writing practice that works for you, and ensuring that it stays effective throughout your studies. Bear in mind at this point that most students in the sciences will write the thesis in either Word or LaTeX and convert it later into a PDF. You should get training as early as possible to make sure you are comfortable working with these pieces of software before you invest a lot of time in something that you'll change later.

We will also encourage you to see writing not simply as the end product of your work, but also as a valuable tool to help you articulate your ideas and refine your thinking along the way. We'll suggest a technique that will show you how to get vital feedback from your early drafts, and help you speed up your whole way of working.

We work with students all the time who are carrying out fascinating research, and who are passionate about what they do – but then hit a wall when it comes to writing about it. Writing is the primary way you will communicate your research to the rest of the scientific community, so let's look at how to make your writing time as productive as possible.

Managing your time

One of the first things to consider when establishing a new writing practice is how much time you actually have to write. Often, when we talk to PhD students in the sciences, we find that writing is, of course, considered important, and that students are very keen to make regular progress and get feedback on their work from their supervisors, but that writing somehow seems to end up near the bottom of the list when it comes to prioritising tasks.

This is understandable; there are many demands on a PhD student's time, and it can be difficult to get to grips with managing those demands. However, if constant awareness that writing is important and should be happening more frequently is combined with a lack of time actually *spent* writing, the result is often a great deal of tension.

Audit your time

The first thing that you need to do when building your new writing practice is to carry out an **audit** of your time, which means **recording and evaluating how much time you already spend on different tasks**. There are a number of apps that can help you track your time. We will provide a list of apps and websites at the end of this book, but at its simplest, an audit could take the form of a spreadsheet with a column for every day of the week and a row for every working hour.

Ongoing external commitments are probably easiest to record first: lab hours, teaching commitments, supervisory meetings, part-time work, etc. Next, fill in commitments that are non-PhD-related, but still essential: commuting, household tasks you carry out, life admin you complete, etc. Lastly, make sure you add non-essential activities that also use your time: going to the gym, socialising, etc.

Now that you have a more complete idea of an average week, take a look at the places in between the filled cells. What happens during that unallotted time right now? Reading? Writing? Administrative tasks?

You might find it useful to complete an audit and then spend the following week monitoring to what extent your time is actually spent the way you imagine. Even something that seems relatively minor, such as 20 minutes spent on admin tasks before and after teaching, can add up to something that takes up more time than you expected, intruding on time you had set aside for other work. You should also take note of how frequently planned time is derailed, and think about what causes this. Do you sit down at your computer to write only to end up responding to emails? Note this down. If you find it hard to monitor this, then tools such as RescueTime, which track what you've actually done with your time and send you a regular report, can help you get an honest picture of your day.

If you're in the early stages of your PhD, you could also try reflecting on how you managed your time during your undergraduate studies and master's studies (if you completed one). Were you happy with your ability to manage your time, or was it something that caused stress? Can you identify any patterns in your behaviour? If you're at the later stages of your PhD, and trying to establish better habits, then look back over your time so far. Were there any particularly productive periods? What facilitated this productiveness? Is this something you could reproduce or foster? Undertaking this kind of analysis now can make for a much happier experience.

Insights from a researcher

As an undergraduate I could stay up all night and churn out an essay just before the deadline. I learned the hard way that a thesis is different. It sounds like a cliché but it's a marathon, not a sprint. Looking back, I now know that on successive days I can expect to produce quality writing for approximately four hours a day. If I could do it again I would plan for four hours of writing and spend the rest of my time making figures/proofreading/planning/compiling references/going for walks.

Now that you have audited and reflected, and have a more realistic picture of how your time is spent, you should have a better idea of how much time you actually spend writing or, at least, how much time you have *available* for writing in any given week.

The next thing to consider is whether you're happy with your rate of progress. If you find this schedule works for you, and you feel that you're advancing as quickly as you'd like to, then you can simply make sure to monitor and maintain the balance you've found.

If, however, you feel that this schedule does not work for you – that you're not getting enough time to write, or that the time you *do* spend writing isn't as productive as you'd like it to be – then take it as a cue to make some changes.

Creating extra time

Can you cut down on your non-PhD commitments? If not, then you need to think carefully about how you might create extra time. The first step to doing this is to reflect on past and present habits, and on your beliefs about how you work.

Think about your undergraduate or master's studies. How much time did you set aside for assignments, and how was that time divided? Did you set aside three days before a deadline and spend every waking moment writing? Did you begin further in advance and write in small blocks throughout the day?

If you realise on reflection that you were a writer who preferred to work for long, uninterrupted stretches, and your audit suggests you could rearrange your time to allow this kind of pattern, then it's something you should consider trying again now. Changes in working habits always have the potential to be initially uncomfortable, but you should persist for at least several days of using your new pattern before making any decision on whether to maintain it or adjust it.

If, on the other hand, this kind of readjustment isn't possible, or if long stretches of writing just didn't work well for you, then **try creating more writing time by committing to more short, concentrated blocks of work instead**. There are two main strategies to consider:

First, **reassess 'dead' or 'lost' time** – time that you might not previously have considered usable. For example, if you have a half-hour train journey, then this is a half-hour that could be spent writing (or doing prep work for writing).

Second, you could also try to **create new blocks that might not have seemed obvious before**: a 10- or 15-minute block in the morning before you check your email, for example. You will be surprised by how much writing you can get done when you have a short, well-defined space of time like this in front of you, especially if you know this is perhaps your only chance to get writing done that morning, or that day. Perhaps you'll also find you get into a flow, and you decide to stay for longer than you originally planned …

If you have sufficient time for writing scheduled throughout your week in a way that suits you, **but you're still unhappy with your progress**, then you need to assess what's actually happening during your writing time.

Are you writing, or are you distracted?

As you might have discovered when you carried out your time audit, it's easy to convince yourself that you're doing one thing when in fact you're doing another. Time spent in front of your computer screen in your office doesn't necessarily translate to working, and time spent relaxing at home can morph into checking emails and thinking about what work tasks you need to do the next day.

It's very easy for email and social media to chip away at the time you think you're spending writing. Checking your email and making academic connections online can both be productive, but they're also an easy way to procrastinate, because you still feel as though you're being productive. There's also evidence to suggest that alternating your attention between writing, which requires a great deal of focus, and superficial tasks like email or checking in on Twitter, can have an overall detrimental effect on your concentration and the quality of the work you produce (Newport, 2016).

If you suspect that this might be a problem for you, or even if you don't, but want to get a clearer picture of how your time is spent, there are numerous trackers available that can monitor in real time how you actually spend your hours when you're working on a computer (we mentioned RescueTime earlier). Choose one and monitor your time over the course of a week.

If you identify that your time and attention are being drained in this way, then you now have the power to act accordingly. You might find that being aware is enough, and you can address the habit easily. Alternatively, if you find it hard to break these habits, there is more free software available that can help. Cold Turkey, for example, will allow you to block specific websites at specific times of day. Another approach might be to use software such as FocusWriter or WriteMonkey, which provide very simple full-screen interfaces, hiding distractions and encouraging you to focus on the task at hand.

It's also worth applying some reflection and analysis to those times of day when distractions seem most powerful. Are you tired? Bored? Trying to avoid a particularly tricky chapter? It's worth thinking about how you can get into a state of mind where you're less likely to be distracted, as well as minimising the distractions themselves.

For a deeper discussion of the value of concentrated work and the impact of distractions, then we recommend reading *Deep Work* by Cal Newport.

Are you writing, or are you editing?

These two tasks are closely related, but still separate. If they're allowed to bleed into each other, this will slow you down considerably. We'll talk in more detail about how to avoid this in the 'Process and product writing' section later in this chapter. For now, however, we would suggest that if you are trying to produce a first draft, going back and tinkering with individual sentences and paragraphs as you write can make progress painfully slow. Try to make sure that you clearly define what you are doing in each writing block. If you are producing a first draft, then get words down on the page. If you are going to edit, then specify that this block will be spent only on editing. Trying to do both at the same time rarely, if ever, goes well.

If you find it difficult to break yourself out of the editing habit, then one of the pieces of software that we mentioned above, FocusWriter, has an option called Focused Text, which fades out what has been typed previously, forcing you to focus on the current sentence.

Are you writing, or are you reading?

Writing is a recursive process. We frequently produce a first draft that points to the need for more reading. That is one of the benefits of a first draft, which we will talk about more later.

There is a tendency, however, for PhD students to get stuck in the well-known 'just one more article' trap, where you tell yourself that you'll feel ready to write after just one more paper. This can often turn into two, or three, or four articles, until your writing time has been entirely consumed by reading instead.

There's a fairly obvious reason for the existence of this trap. As we discussed back in Chapter 1, undertaking a PhD is a huge transition – taking you from student to researcher. In the early stages, it's natural to lack confidence, and to feel like you need to read more and more before you feel secure enough to make a statement, or offer an observation, or make a critical comment about someone else's work.

The result of this trap is usually a blank document, several open tabs with various journal articles and pages of notes, or annotated articles which, while they might be useful, will need to be carefully considered before they can be factored into your current piece of work, or which might actually have stalled your progress by making you second-guess the whole direction of the section/chapter.

The recommendation here is the same as it is with writing vs. editing: **decide whether you are reading or writing.** If you're writing a first draft and you encounter a gap in what you're trying to write, or an area you feel would be strengthened with more reading on the topic, then make a note of this in the document and move straight on to the next section that you *can* write. Alternatively, if you find that there is no way to make progress without substantially more reading, then cancel the writing session and set the time aside to focus only on reading instead.

Writing environments

The next factor to consider in building your writing practice is environment. Where do you do your writing? From our experience, students vary hugely in where they like to write. Some like to work in their office, if they have one. Others prefer the library, where they have other people around them, and ambient sound, but are unlikely to be interrupted by someone they know. Some prefer coffee shops for the same reason. Other students prefer to work at home, where they can have more control over their environment and minimise distractions. Others, alternatively, find that home working offers too many distractions. Working on public transport is popular, too.

If possible, it's obvious that you should try to ensure that your writing time is spent in the environment that's most conducive to progress.

Insights from a journal editor

I started writing my PhD thesis in the lab while simultaneously doing practical work. This part-time approach worked well for some time (particularly as I was able to quickly fill the gaps in the data that became apparent during the first phases of writing), but I eventually understood that I needed long stretches of uninterrupted time to write and concentrate on the complexity of the thesis.

I decided to move to the local library for a better writing environment and postponed any practical work that came up for when the first draft of the thesis was completed. I was much more productive and was able to write long stretches of the thesis there.

Towards the end of the draft stage I required a change in scenery and began writing at home. Home was a more isolated place, which I enjoyed at that late stage, but was full of potential distractions. I defined 'work hours' and 'play hours' – I felt I needed substantial downtime to keep my focus sharp after so many weeks of writing. Following this schedule required strong discipline!

Dr João H. Duarte, PhD (Senior Editor, *Nature Biomedical Engineering*)

The example above clearly demonstrates the value of regularly reflecting on your habits. It's easy to become too fixed in our notions of what works best. Try new places, just to see how you get on. You might be surprised to find that the library works better than you might have suspected. Compare work rate, too. You might find that you're more productive in a new location, especially if you've come to associate your usual writing space with frustration and lack of progress. Starting in a clean location can be a good way to kickstart a new writing practice.

If, for some reason, your usual writing space becomes unavailable, or you have a schedule change which means that you can't access it anymore, then it's worth trying to reflect on why that space worked for you, and whether you can replicate those conditions elsewhere.

For example, perhaps working in the library on campus puts you in a 'work' state of mind, since you are on campus, but you are still outside the environment of your department, where you're more likely to be interrupted by students, or to be aware that your supervisor is only a few doors away, or to feel pressure to make conversation with your officemates, or worry about whether they're making more progress than you... If that library became unavailable to you for some time, then would a computing lab serve the same purpose, or maybe even a common room, or a study space that's more usually used by undergraduates, or by students from another discipline?

The type of noise generated in these different spaces can also have an impact on your ability to concentrate. Some students need complete silence, while others find that silence is more likely to make their mind wander and latch onto distractions, preferring ambient sound instead. Many people like to listen to music while they work, but often prefer music without lyrics, which can be distracting when you're trying to articulate your own ideas and think about how things should best be worded. You'll find plenty of research into what type of music works best for study, and which offers up explanations about why the 'best' type is the best. We haven't found any articles that conclusively show a certain style of music is objectively better than any other for all types of people, though, so we'll stop short of endorsing any particular type of music and encourage you to experiment for yourself, and not to let any small-scale study try to dissuade you from believing your own observations about *what works for you*.

Again, the key is to create, to the best of your abilities, the whole set of environmental circumstances that best suits you. If you're forced to work for a while in a more disruptive environment, and find it hard to work to music, then there are many good ambient sound generators available online. One free resource is mynoise.net, which contains several sound environments that you can then adjust to suit you. Another good recommendation we've had from students who are distracted by hearing other words, but who also prefer not to try to block them out with *music* while they work, is to listen to radio stations or podcasts in a foreign language.

Motivation

Spending time considering what actually makes you sit down and write can also be helpful in analysing your current habits and thinking about how to build motivators into your writing practice. When we've asked students to think about what motivates them to write, answers to this question can be roughly sorted into positive and negative motivators.

On the positive side, we tend to want to write when we feel confident about our knowledge. Having something interesting that we want to communicate, either to our supervisor(s) or to a wider audience, is another motivator, as well as being inspired by other work that we've read. Having good data is another common response to this question. Being keen to get feedback on progress is also a frequent answer.

We sometimes hear the rather vague 'being in the mood to write', but on further discussion, this is usually simply the unanalysed combination of all the above factors: feeling inspired by a good article, confident in our own understanding, and clear about what we want to say.

On the rather more negative side, students will sometimes say that guilt and/or fear motivate them to write. While it's possible to take an 'ends justify the means' approach to this – that the motivators don't matter as long as there's an end product – it's not the most positive way to work.

It's important to differentiate here between different kinds of fear and guilt. Fear and guilt over the prospect of missing a deadline and using the resulting stress to motivate work is fairly common. If both emotions are reasonably mild, and prompt action, then you need to reflect on whether you find this manageable, and to what extent you want this to be part of your routine.

If, however, fear and guilt are prompting long spells of paralysis before frantic last-minute writing, and are making writing an anxiety-laden task, then there are possibly deeper issues to address. Writing can be a revealing task, and some students put it off because they are simply scared that, when they write, they'll find that they know less than they thought they did.

For the time being, you should simply reflect that perhaps you've found that these motivators are actually not as truly motivational as you suggested, and you might look to the 'Roadblocks' chapter for more information on how to address these issues.

Keeping a thesis journal

You've seen from the preceding discussion that people's motivators are complex, and often actually have relatively little to do with their perceived skill in writing. Your willingness (or otherwise) to write is often more symptomatic of how you feel about your wider progress in the PhD. Being aware of this will give you the ability to analyse your current situation and tackle these issues, and is infinitely more useful than the vague label of 'writer's block', which is both misleading and unhelpful. **The key tool to diagnose stalls in your progress is reflection (which you can think of as self-analysis).**

Reflecting on your writing practice through the lens of the factors we discussed above can help give you a much clearer picture of who you are as a writer. We would recommend reflecting regularly throughout the PhD, as your needs and situation are likely to change, and being aware of when this happens will allow you to adjust your practice accordingly.

To enable you to do this, as well as reflect on the process more generally, consider keeping a thesis journal. This can take any format you choose. There are several journalling apps, available across all platforms with varying levels of sophistication, which can give you daily reminders to keep using them. Some you could consider include Day One, Penzu, Journaly and Diarium (all with free options). Alternatively, you might prefer a plain notepad and pen, especially if you find your phone to be a source of distraction.

Equally, the degree of depth and formality is entirely up to you. You can write a detailed page on how you've decided to change your approach to a specific problem in your work, or you can simply note that your work rate slowed in November due to taking on too many commitments, or that the weather made you feel sluggish and unmotivated.

What's important is that you are constantly reflecting on the experience, which will enable you to spot patterns in your behaviour and to subsequently curtail or cultivate situations which have an impact on your writing progress.

A further benefit is that **if the act of writing, until this point, has become associated with stress and guilt in your mind, then getting into the habit of writing every day reframes it**, and makes it an everyday activity. Here's a piece of advice from someone who has recently completed their PhD.

Insights from a researcher

Practise writing. Anything. All the time. The more you write, the better you get. Don't leave it until the end. I always had a picture of the shape of the thesis in my head. Of course, I didn't know the content until I did the research, but I always knew that the thesis existed. This helped a lot when I was struggling, or at a stuck point.

As this shows, writing for *yourself* (as opposed to your supervisor, or a peer reviewer, or a progress review board…) allows you to reclaim the act of writing as a valuable means of analysing your own progress and working through tricky patches, as opposed to solely perceiving it as a means of external assessment.

Process and product writing

It's important to remember that writing your PhD thesis will often be difficult. This doesn't necessarily point to a lack of ability in writing. You are writing at the highest academic level. Your work, at this point, involves synthesising complex ideas into a coherent context, clearly articulating the contribution your work will make, communicating complex information and finding a fine balance between concision and depth. This is not an easy task, no matter how practised the writer is, or how familiar they are with their subject area. Writing will often be hard, and might often be frustrating. This does not mean that writing is not going well.

We talked in the previous section about some factors that can both slow your progress and make your confidence in your writing waver. We discussed how easy it is for the lines between writing and editing to blur, as well as how fear over the quality of your work can slow your progress.

One model that we think would help to address these issues and develop a better way of writing is that of process and product writing. This theory, first defined by Murray (1972), and further developed and refined by theorists such as Flower and Hayes (1981), comes from the field of compositional studies.

At the moment, you might, if you are similar to many of the students we work with, produce one working draft for any given task. After this first draft has been produced, you work within it, editing it and reshaping it until it is the finished version that you will submit to your supervisor.

Process and product writing would instead suggest that the first draft and final draft are two completely different types of writing.

Process draft	Product draft
Record of thinking	Presentation of fully developed ideas
Disorganised and rambling	Coherently and logically structured
Writer-focused	Reader-focused
Imprecise word choice	Incisive word choice
Repetitive	Concise
Uncertain, possibly over-reliant on supporting material	Confident in own work and judicious in use of supporting material
Several gaps where more reading is needed	Comprehensive and fully contextualised
Thinking	Explaining

The process draft

You might recognise several features of the left column, the process draft, as characteristic of your first drafts. **First drafts are many people's least favourite part of writing**, no matter how much writing they do, how proficient they are and how many papers they've published. They're often difficult to write, and not of the standard you would hope to produce.

When you think about it, there's a good reason for that. The first draft is often the *hardest* part of writing: **your first attempt at articulating your thoughts on the page, taking complex and dense ideas, and expressing them in a sufficiently sophisticated series of words**. This is by no means an easy task, especially where you might well be trying to advance a completely new idea which no one else has ever expressed before, or detailing an especially complicated process.

Additionally, at the time of writing the first draft, **you may not yet have a solid grasp of the material as you would like, or fully understand how certain ideas are connected** – especially if you are writing at an early point in your PhD. You might be unsure of how much context you need to provide, or of how to present your work as a logical progression of what has come before. If you're writing the literature review, then you're probably still trying to digest a huge amount of new information, as well as thinking about how you want to respond to it.

All of these factors manifest themselves in your first draft. Instead of the concise, coherent style you would see in a finished paper, **you might notice that some concepts aren't well described, or that your argument loses coherence at points, or that it isn't entirely clear how certain ideas are connected**. You might be overly reliant on quotations from existing studies.

It's important to remember that this tends to be what all first drafts look like, no matter how experienced the writer is, and how advanced they are in their field. These first drafts are the drafts Murray described as 'process drafts'. Process drafts do have a hugely important role to play in the writing process, but they're not the bulk of the finished version of the document that you will hand to your supervisor for feedback.

Writing this first draft can become a much less stressful experience, and also become much more useful to you, if you think of it not as the draft you will eventually hand in to your supervisor, but instead as a process draft – a draft with a very specific role. **Process drafts are not a reflection of your ability as a writer**. Process drafts are instead a reflection of your current understanding of your project and your ability to articulate that understanding through the written word *at the current moment in time*.

Let's discuss how you can use the process draft to your advantage.

How to benefit from the process draft

Once you understand the nature of the process draft, you can understand that process drafts play a vital role in your thinking and in your writing. Problems in this draft should not be seen as failings and a source of frustration, but should instead direct you in exactly what steps to take to produce an improved second draft. For example:

- **Gaps in the text** where progress stalled can point to a need for extra reading to fully grasp complex ideas and consolidate your understanding.
- **Very short paragraphs** point to ideas that need further development or ideas that would fit better into another section of the chapter. Paragraphs

that are very short can often be absorbed into a larger paragraph, so you might have to think about how the information contained could perhaps be integrated with a larger idea or block of information elsewhere.
- **Paragraphs that are far too long** and lose focus often have too broad a central idea. If you read the first line of the paragraph, the topic sentence, then you are likely to find that it is not specific enough to produce a coherent paragraph which puts forward one main point. This might point to the need to break some of your ideas down in order to examine them more effectively.

You should also highlight the content you're happy with, and those aspects of the paper that you think work well. If you've hit on a particularly effective way to phrase something, or if detailing something in writing has led to a fresh perspective or improved understanding, then take note of this. Process writing reflects everything, and that includes good ideas and solid understanding, as well as areas where more work is needed.

As soon as you accept that **the process draft is for your eyes only**, and that its imperfections are its strength, you will also find that you can produce this first draft much more quickly. You can note in the text, if you like, places where you faltered or where you can now see a gap, but you ideally want to aim to simply get words on the page. Instead of seeing drafting as a strictly linear process, you can now see that writing is a recursive process, and often looks much more like this:

> Reading – Note-Taking – First Draft – More Reading – Second Draft – Possible Third Draft – Final Draft

The other great advantage of acknowledging that the process draft is not the draft that will be handed in to your supervisor(s) is that it frees you to write without anxiety, and without interrupting its production to go back and try to edit sentences and paragraphs. No one is going to read this draft other than you. **It doesn't matter if the tone needs work, or if it lacks concision, or if the structure is untidy**. This draft is not going to be assessed. Its sole purpose is to let you see exactly where your understanding is, and what further work is needed.

After you've undertaken the additional work the process draft pointed you towards, you should go ahead and produce a second process draft. The same essential principle still applies: this draft is a tool for you, not a piece of work for your supervisor(s) to assess. However, with the information you've taken from the first process draft, you should be able to be a little more targeted and comprehensive in your writing.

After you've produced this second draft, you undertake exactly the same process as you did after the first process draft. You might find, at this point, that you still need to go back and do more reading, and then produce a third draft. Alternatively, after highlighting key content and strong points, you might feel that you know exactly *what* you want to say. It's now time for you to produce the product draft, which is about *how* you say it.

The product draft

If you think back to the table on page 22, you might remember some of the key attributes of this next draft: it's well structured, concise and coherent; the reader is given exactly as much context as they need to follow the work; supporting evidence is used judiciously.

As you're now aware, the product draft benefits from the work invested in the preceding process drafts. You know exactly what you want to say. The next step is to put yourself in your readers' shoes and think about the best way to communicate this information to them.

We suggest asking yourself the following questions before proceeding with your product draft.

1. How much context is needed for the reader to understand this chapter? Is there any vocabulary that might need to be defined? Are there any theories which might need to be clearly explained in order for what follows to make sense?
2. Are there any sections where I need to guide the reader through a complex process or where I connect two ideas in a new way?
3. Where do I need to make sure the most impressive parts of my work are foregrounded? Where can I make my contribution evident?

You can now consider the best way to address these questions through writing.

For the first set of questions – which are largely to do with providing sufficient context and explanation – you might find it helpful to imagine not that you're writing for your supervisor, but that you are writing for a first-year PhD student studying the same subject as you.

There's a good reason for this. When we write for our supervisors, we tend to under-write. Since our supervisors are more knowledgeable than we are, there's a tendency to assume knowledge and understanding on their part, and skimp on explanation. In practical terms, this means that terminology might not be clearly defined and ideas might be under-explained. If you write for a first-year-level PhD student, however, then you're much more likely to provide the appropriate level of explanation.

This method holds true for writing for publication, as well as the PhD. Think about writing for a knowledgeable reader who, nevertheless, may not be fully familiar with your topic. Ask yourself what terminology might need to be defined or what previous studies might have to be discussed in order for someone to understand the novelty of your own work.

For the second set of questions – which are more to do with guiding the reader through tricky parts of the text – you might want to carefully consider how you use connectives. Connectives are often overlooked, but they play a vital role in our writing. They include words such as:

> consequently, because, previously, next, however, alternatively, in brief, to conclude, notably, particularly, crucially

We'll look at these connectives and the role they play in more detail in Chapter 10, which deals with paragraphing. For the time being, it is enough to understand that they're especially helpful in sections where the reader might require extra guidance, because they tell the reader exactly how one piece of information is related to the text, and therefore how they should interpret it. They guide the reader through complex sections and help to mitigate the risk of misinterpretation or confusion.

You might also want to think very carefully about whether there is a logical progression from one paragraph to the next. The information on reverse outlining on page 138 might be especially useful here, allowing you to see the skeleton of your chapter, and pinpoint any problems.

For the third set of questions, you might want to carefully consider your structural choices. This will be examined in great depth in Chapters 9 and 10, which deal with sentences and paragraphs, respectively. For now, however, it is enough to ask yourself where you have placed key pieces of information in the draft. Have you buried a particularly astute observation in the middle of a long, multi-clause sentence? You run the risk of the reader glossing over it without fully recognising how good it is. Have you left a novel idea languishing in the penultimate sentence of a long paragraph? Again, burying it in this way means that it might not receive the recognition you would like, and inadvertently downplays its importance.

Benefits of the process and product approach

The process and product approach might initially seem like it involves more work. However, if you think back to the last piece of work you produced, and how difficult it was to twist and mould your first draft into a finished version that you were happy to submit, then you can see how this method might actually speed up the process. Instead of taking your first draft and trying to force it to perform a task it was never designed for, you can instead start fresh each time, identifying and addressing gaps and weaknesses as you find them.

On top of this, **this method removes a lot of the anxiety around producing the first draft**. Nobody but you will read it, so there's no need to worry unduly about how it appears right now. This can also go some way to tackling perfectionistic tendencies. The process draft's job is not to be perfect.

This new approach to writing can also be factored into your writing practice. Some students, after adopting this method, have, for example, chosen specific locations for the production of the process and product drafts: writing the process drafts at home, but producing product drafts in their office. This reinforces the notion that **they are two different types of writing**, and helps you to avoid slipping back into the habit of editing the process draft.

However you choose to shape your writing practice, the key principles to take from this chapter are the power of reflection in helping you to address issues and maximise your efficiency, and – most importantly – the recognition that writing is a means of thinking and of refining your ideas – a powerful tool at your disposal, and not simply a means of external assessment of your abilities as a researcher.

Reflecting on your writing practice, assessing your habits, introducing new working routines – these all require time and effort. Bear in mind that a productive writing practice not only makes your life easier during the PhD, but will help ensure you produce a piece of work that you can feel justifiably proud of at the end of the process.

Insights from a journal editor

Writing a PhD thesis is definitely a tiring process, and, probably inevitably, there comes a time when you want it to be over. But I think it's important to remember the bigger picture – the monograph you're writing is your contribution to science, synthesising a few years of hard work, of which you should be proud. Although now 'just' a proud item in my bookshelf at home, I regularly picked up the hard copy of my thesis during my postdoctoral years to look something up – finding errors or typos once in a while – to the point that it has almost fallen apart.

Bart Verberck (Regional Executive Editor, *Nature Research*)

CHAPTER 3 # Refining and Articulating Your Research Question

Scientific investigation is usually taught as being based on a hypothesis. The main aim of this chapter is to show you that it's not crucial to start with a hypothesis right from the very beginning, nor even to have an overarching one *at all, ever*. We'll explain the difference between a **hypothesis** and a **research question**, and we'll show you the merits of **hypothesis-led** research and compare it with the merits of **hypothesis-less** research.

This chapter will also put those concepts into some context, given that the type of scientific research conducted around the world today is very different from the research carried out by the pioneers from history who set out the original rules of science, whose style of work still forms the basis of most people's perceptions. Large datasets are publicly available for interrogation after the fact; researchers are not just investigating the fundamentals of how things work, but developing more and more applied solutions to problems; abstract and theoretical sciences don't always pertain to immediately testable concepts – the list of reasons to operate without a hypothesis is long. Of course, we're not saying you should necessarily *aim* to proceed without an overarching hypothesis, but we'll show you how to design the type of project that doesn't have a well-developed one at the start.

You'll probably find it useful to return to the latter part of this chapter after your first several months of research (or beyond). It is probably more common than not for project plans to change, and so you'll almost certainly find yourself refining and finalising the research question(s) you'll present in your thesis months – or even years – after your PhD begins.

The (un)importance of starting with a defined project plan

There are two aspects to consider when looking at PhD studentship positions: what the project will involve, and where the money will come from. These are closely related, as the decisions about the broad research direction are often – but not always – shaped at the time of applying for the money. Depending on the PhD you apply for, you may be required to think up your own project, or you may be joining a project that is already underway.

If a PhD supervisor has already secured funding and is advertising the position, then it's very likely that the project's goals will have been written for you already. In this case, it would be up to you to take those plans and start to act upon them. Although your supervisor will have read the literature to inform their plan, you may find that there has been new literature published in the time since they submitted their grant proposal. You might then have to make significant refinements to their plan, in which case the first steps you planned for your research project may have to change. For example, someone may have recently published something that answers part of the question you intended to ask. (For more on being 'scooped' and how to deal with it, see Chapter 4, p. 52.) If, on the other hand, you've secured a stipend to cover your living costs and you now need to find a supervisor to take you on, the challenge of writing a research proposal will probably become more of your own responsibility, at least to begin with.

Research questions vs. hypotheses

In short: a hypothesis is a statement that can be tested; a research question is not.

Research questions, in contrast to hypotheses, aren't testable statements of fact. They should be phrased as questions, with question marks at the end.

For example, if you'd simply like to test the impact of carrying out a procedure, a research question might be the more appropriate way to define your project's initial aims.

> *Do proteins X, Y and Z from dog saliva have an impact on the speed of human wound healing?*
>
> *Can aeroplane wing stability be improved by the application of [a piece of knowledge from a related piece of existing work]?*

This allows you the freedom to try to apply the knowledge in various ways; ways that might not occur to you until you are, for example, in the lab holding a model of an aeroplane wing and considering the actual ways in which you could reasonably modify it with the tools available to you at the time.

Another reason for leading with an open-ended research question might be that you have a large dataset and you want to interrogate it to check for new connections. For example, anonymised genetic data from various populations is available on a large scale from commercial companies and national health boards. With the appropriate ethical application clearance, you could interrogate these repositories of genetic markers to search for correlations between the genes and various health outcomes.

You might find a relationship between three different genes that have a combined effect on the likelihood of developing one certain health problem. In this case, you would have had no plausible way to formulate a *hypothesis* in advance, but as a [disease X] researcher, interrogating the data would be a perfectly valid scientific way to approach a more open *research question* like this:

> *Do any relationships exist between genetic markers in a population of Latin American women and their likelihood of developing [disease X] between the ages of 25 and 50?*

Similarly, research questions are more suitable for complex problems with huge internal variability. For example, a researcher in the field of public health may have read that family-based educational programmes and school-based educational programmes have both had moderate success in improving obesity rates in adolescents in various pilot studies. They might reasonably conclude that a combination of the two approaches might have a better effect than either one on its own, but people make hugely variable samples. It would be very difficult to formulate a *hypothesis* about how much weight loss could be achieved, and difficult even to know how to quantify it. (Average weight lost per person? Number of adolescents who achieved significant weight loss? Length of time each adolescent's weight loss persisted after the end of the intervention?) Instead, it would be more sensible to start with a *research question* that encompasses all of these possibilities, and more:

> *Does providing healthy diet and lifestyle information to overweight adolescents through a combination of home and school interventions at a national level lead to a meaningful reduction in obesity levels in the UK?*

Hypotheses, on the other hand, are less open-ended than research questions, and should be written in the form of testable statements of truth.

> *Applying chemical X to sample Y will result in a reduction of 50% in measurable quantity A.*

There may be more than one plausible hypothesis; in this example, '50%' could only be one possible prediction based on various models of how the system should perform. Formulating several hypotheses before you start is not just a useful intellectual exercise, but it can help you to predict what possible next steps you might need to take in your experiments. For example, if the experiment was run once and 50% *wasn't* achieved, would you consider that to be a failure in your lab set-up, or would you be satisfied to record the number and move on? What if you could have changed something and 50% *was* then achieved? Creating sub-hypotheses about how various elements of the process might work can help clarify your project plans. This can help you prepare the relevant materials to investigate these options in advance, or to pre-emptively book extra time on the relevant pieces of instrumentation.

In addition, the **null hypothesis (H_0)** is the one that predicts your intervention will have no impact. It is a statistical tool that functions as a sort of threshold for the impact of your intervention. If the variations in measurements can be attributed to random fluctuation, the null hypothesis holds. If the variations *can't* be explained by the expected natural variation, scientists talk of 'rejecting the null hypothesis' – that is, their intervention caused the result. (We won't go any further into statistics here, as it is too broad a field in its own right to adequately introduce many meaningful concepts within the remit of this book. We recommend that you consult your university doctoral training resources or your university library for more specific guidance.)

Hypotheses should be as specific as possible. This means including – as much as you reasonably can – the defining characteristics of the people or the things under examination (the 'population group'), the defining details of the experimental intervention (e.g. concentrations, durations, temperatures, etc.), and specifics about the outcome measure (or measures) that you're interested in.

H_0: Adding chemical X will have no impact on outcome A.

H_1: Adding chemical X will cause a 50% reduction in outcome A.

H_2: Adding chemical X will cause a 75% reduction in outcome A.

Hypotheses can never be proven (except in very specific circumstances – see below). They can only ever be **supported** or **disproven**. Part of operating by the scientific method is avoiding claims that cannot be backed up by empirical evidence. 'Proof' is a protected term in science, and if something is 'proven' then it must be literally, absolutely true in all cases at all times to be awarded that status.

Take the structure of this very simple hypothesis:

Mixing equal volumes of an acid of pH 6 and an alkaline solution of pH 8 will result in complete neutralisation (i.e. pH 7) within 10 seconds.

It's a good hypothesis in that it's **testable**. It may also be true enough for all practical laboratory purposes, but there may be molecules of the reactants that haven't been converted to products by the end of the 10 seconds. These may not be detectable by our techniques today, but there's a very real possibility that detection might be possible in the future. We would therefore avoid saying that the hypothesis had been *proven*, as we can already see limitations that prevent us from making such a confident assertion.

Consider also that there's vagueness in the phrasing of the hypothesis itself that prevents it from being applied in all situations, or prevents the results being interpreted equivalently by all readers. In order for all of the products to react with each other, there must be sufficient mixing after they're poured into the same vessel. With a small volume of reactants in a desktop beaker, 10 seconds might be a reasonable timescale, especially if the solution was to be stirred. Mixing equal volumes of acid and alkali in an industrial tank, on the other hand, is likely to take much longer to reach equilibrium. It's therefore not possible to say we've *proven* this hypothesis when we can already see that the phrasing is vague enough to include situations that would actually behave differently from our experimental set-up in the lab.

Finally, the laws of probability simply interfere with our ability to be certain. It's highly *improbable* that a small volume of the two solutions would fail to mix thoroughly in a small benchtop beaker, but that doesn't mean that it's *impossible*. If the experiment was to be repeated thousands of times, there could conceivably be a situation where most of the acidic molecules would partition on one side of the beaker and most of the alkaline ones on the other side, and that they would take a long time to mix.

We therefore can't say that the hypothesis has been proven *in every possible situation*. This means it's not technically a statement of pure, objective fact, so we should avoid saying it's been proven at all.

In each of these cases, we can only say that the hypothesis has been *supported* (or, alternatively, *not yet disproven*).

The exception to this might be when a hypothesis defines very narrow experimental conditions, for example to a specific period in time:

> *[Radical child support policy change X] will decrease the number of mis-filed applications for child support by 10% within the first year.*

In this case, if it turned out that wrongly filed applications did go down by 10%, then we could say that the hypothesis *was indeed proven* to be accurate, as the chance for it to have been disproven would have passed. Note that this is different from saying something like:

> *Shining a torch into a dark pond will stimulate an aversion response that increases the proportion of fish hiding under vegetation by 10% within the first minute.*

Although this animal behaviour experiment is also time-limited, the number of opportunities to repeat the experiment are *un*-limited. In other words, it is not linked to a specific point in time, it could be repeated, and it could therefore conceivably be disproven at some point in the future.

Terminology note: hypothesis vs. theory

Theory is a protected word in academia. In common usage, it simply means a prediction. In academic circles, it means a collection of observations and conclusions that combine to create a sub-field of study. Evolutionary theory, quantum theory, atomic theory, molecular orbital theory – these are all examples of well-established scientific theories, but none of them are equivalent to hypotheses. They are frameworks of existing knowledge that allow us to synthesise and develop further assertions and testable predictions on each of those topics.

Which one should you choose when you write your thesis?

Hypotheses are well-suited to situations where a rich enough set of background information exists to make predictions possible. They are also common in quantitative research, where it is easier to create defined numerical cut-offs to go into your predictions.

Hypotheses don't suit every type of research, though. The more complex a system is, that is, the more variables it has or the more stochasticity ('randomness') is inherent in it, the more difficult it can be to formulate a meaningful hypothesis. Hypotheses are also easier to write when you have a narrowly defined scientific problem to tackle.

Research questions, on the other hand, are more suitable when you have a general goal in mind but you're less sure of the best way to achieve it.

You needn't actually pick one over the other, though. As you'll see below, a good project that begins with a main research question might then be broken down into several sub-sections, each of which is led by a testable hypothesis.

Why is all of this important?

On a basic level, it's important for your own credibility to be able to use the correct terminology. On a more practical level, a PhD progress committee might ask for your working hypotheses or your research question, and it's important that you can supply the right type of information.

How do they relate to each other?

Having one doesn't mean you can't have the other.

Perhaps you have a research question that defines your overarching PhD project. It's possible that your next steps once you get started on your research would be to break it down into sub-experiments, each of which has a testable hypothesis associated with it.

In this way, the relationship between a research question and a set of hypotheses is sometimes similar to the relationship between a **methodology**, which describes the general type of research, and individual **methods**, which provide details of what you did at a level that someone else could use to repeat your processes.

Don't expect that your project's goals will stay unchanged

This is a concept that may be very familiar to you if you've completed an undergraduate or a master's dissertation. Initial plans might have to be changed because you find that the literature or the raw material to analyse simply don't exist in the right quantity or the right state to be useful. Perhaps you find that the answer was massively more simple than you originally expected, and now you need to find something else to fill your PhD with. On the other hand, perhaps the answer is much more complicated and unexpected. None of these situations is rare.

Don't worry, though, as nobody will be surprised. Finding that your results don't turn out as expected is probably the most common reason for updating a research question. A PhD isn't a test of your ability to find the answer to your original question, though. It is a training programme designed to train you in how to design experiments, interpret results, think about how they fit with the wider field of knowledge, and then decide what to do next. Scientific experimentation is an inherently cyclical process, so you should expect to be re-designing your plans a lot (see Figure 1).

A slightly different reason for changing your project plan is that you might find someone else has answered the same research question you were working on (in other words, you've been 'scooped' – for more on this, see Chapter 4, p. 52).

Figure 1: The Cyclical Nature of the Research Process

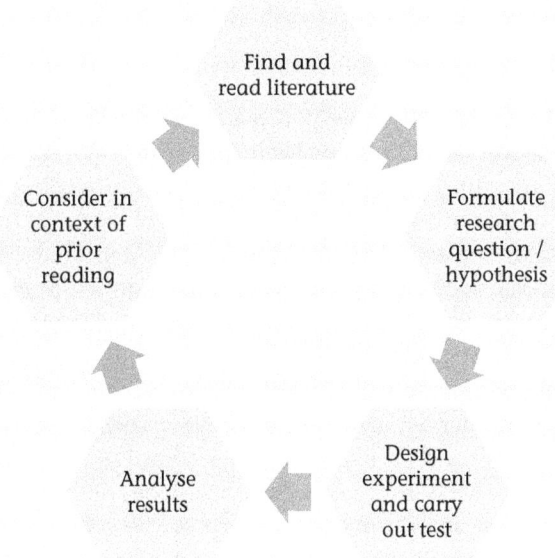

Whether you're scooped or not, and whether your experiments go well or not, the nature of working at the cutting edge of human knowledge is that things are unpredictable. Your supervisor won't be disappointed in you if things don't go as planned, but part of the personal development expected of you throughout your PhD is the ability to become an independent researcher. What they *will* want from you is some suggestion on what you think you should do with the research direction. Hopefully this chapter has given you the tools and the framework to formulate your own research questions.

CHAPTER **4** **Finding Literature**

This chapter is intended to give you a very practical guide on how to carry out a thorough and rigorous literature search.

Going from your undergraduate dissertation, through the process of a master's, and on to writing a PhD thesis, involves learning to handle an order of magnitude more references than you used in the former (that is, your total number will involve using an extra digit – '35' might now be '250'; a typical scientific PhD thesis might have 200–300 references), and so it's important that you have a system for handling these. You might read some of them once in your initial literature review phase, and then once several years later when you write up your thesis. If you don't have a system to organise and take notes on them, your sources might start to feel like a messy collection of texts that you avoid going back to again, but which you know you'll have to one day. We'll go into more detail on organisation strategies in the next chapter, but having a good search strategy in the first place will help you lay the groundwork for keeping a systematic collection of your sources.

This chapter will also feature advice from the experts on information searching and curation – university librarians. These aren't the people who help keep the library operating smoothly by putting the books back on the shelves and helping users with queries; these are the people who *run* the library. They communicate regularly with publishers of academic literature, and decide what books are worth buying and what databases and journals are worth subscribing to. They have the most up-to-date knowledge of the tools available to researchers for accessing the world of academic publishing.

If you've never made use of your own university's subject librarians, check your institution's website to find out what they can do for you. You might find they have web pages or other guides dedicated to helping you navigate your own local set of subscription services. You'll almost certainly be able to email them to ask for advice. You might even find that they have 1:1 appointments or workshops where you can bring your work along for a personalised session on how to find and collate relevant resources from the literature.

Insights from a researcher

I started with an extended literature review at the beginning that took about three months. Sometimes I thought I should get started on experiments rather than read. In hindsight I really benefited from this initial summary in the final stage. I also used referencing software and kept PDF copies of all relevant papers organised in categories.

As I am not a 100% digital person I printed off a lot of papers and added notes by hand. Now, I would rather use a tablet (or the like) to mark up and comment directly in the PDFs. It makes searching for relevant sections much easier later.

Finding a good, authoritative, wide-reaching (yet subject-focused) database

The advantage of using a database rather than a journal's own search page is that you get to interrogate many, many different journals with just one search. There are tens, if not hundreds, of good academic databases available online at the time of writing. Each one is tailored for a specific set of subject disciplines, but many overlap with each other in the materials they cover, or 'index'. As well as indexing different sources, different databases will offer different functionality.

It's important to distinguish between *academic* databases and more general ones. At the time of writing, **Google Scholar** is a popular database that offers an academic search service, but **we would strongly advise against using it**. Bear in mind that **'academic' does not equal 'scientific'**, and the details of which sources Google Scholar pulls on and how it carries out its searches are not as transparent as they are for other providers.

You should start by finding databases that are specific to your field. There's a greater chance that these will cover the sources that you need to find, and that they will be powerful enough with specialised search tools for you to be able to conduct a rigorous, repeatable, reliable search for which you define all of the parameters yourself.

We'll first explain what exactly a database does for you, then we'll give you some criteria to help you judge which ones will be best for your project.

How does an academic database work?

Just like any online search engine, an academic database automatically collects information about the contents of a range of sources. This is referred to as 'indexing'. In academia, these indexed materials could be journal articles, reviews, patents, books and more niche sub-areas of these information types, such as how many times an article has been cited in other works.

Academic databases exist for all subject areas, but it's generally the case that scientific databases offer much greater functionality than those covering the arts and humanities (so if you're moving from an undergraduate course in something like history into a PhD in something like archaeology, you might find that searching these new databases means you need to develop a very different type of skill).

Using the database can be as simple as entering a single keyword, or as complex as entering a list of keywords connected by various commands ('operators') and applying a host of filters to narrow down the results. We'll explain more about these below.

How do you choose the best database(s)?

The precise details of how each commercial database works can be a bit of a trade secret, as each one wants to be able to retain a competitive edge so that consumers continue to choose to use them. The two most important factors in choosing the databases you'll use – and these will be where you turn for the whole duration of your PhD, so it's worth investing some time thinking about these – are **indexing coverage** and **search functionality**.

Indexing coverage

Generally speaking, major databases will provide a list of which sources they index so that you can decide how relevant they are to you and your research. You might even find that your subject librarians have already listed all of the databases that you subscribe to and organised them by discipline already. Check your library webpages or contact your librarians to see if this is the case.

If you're not sure whether a database is going to be optimal for you, investigate their home pages to see if they tell you which journals etc. they index. If you can spot some obvious gaps, perhaps you'll choose to use that database in combination with another one, or perhaps you'll decide there are simply others that are better. Don't worry if you can't spot these gaps immediately – it'll take some time before you become familiar enough with your field to know what the most central journals for your subject are.

There's no guarantee that there will be one single database with the best list of indexed sources for your research project. It's good practice to think about a balance between two important concepts in information retrieval, each of which is something of a contrast to the other: **sensitivity** and **specificity**.

> **Sensitivity:** *The ability to find all of the relevant source materials that are relevant to your search term.*
>
> **Specificity:** *The ability to eliminate search results that are only very slightly related and not of interest to your project.*

It's important to have **sensitivity** because you don't want to miss an article that contains the answer to the research question you're about to invest time and money in.

It's important to have **specificity** because you don't want to spend unnecessary hours reading literature from subject areas that use the same *words* as you, but which mean a different thing in a different subject context.

We'll return to these concepts later in the section on 'Constructing a rigorous search' on p. 43). For now, it means you should think about choosing an appropriate range of databases to make sure you have enough

sensitivity that you don't miss any important journals or sub-fields within your research area, but not so many that you sacrifice **specificity** by returning results that are actually from completely outside of your field.

Functionality

Any good database will let you **filter** and **sort** your results on a range of criteria.

> *Filtering:* Narrowing your search results on the basis of the criteria you select.
>
> *Sorting:* Deciding on the order in which the results are shown to you.

Filtering options might be available when you're typing in your search terms at the start, or after you've seen the results – or both.

For a good range of criteria, search for databases that let you filter/sort based on:

- publication year
- source type (article, review, book, government policy document, etc.)
- language
- citation count (the number of times a source has been referenced in someone else's work)
- research area
- whether open access or not.

Really good databases will let you create an account and save the results of your search so that you can return to them later, and they'll also let you export the citation information for the results you want to use. These citation details will be very useful when it comes to curating your reference list, as it removes a lot of the drudgery of typing things out by hand. For more detail on this stage, see Chapter 5 – 'Organising and Keeping Track of the Literature'.

Finally, the best databases will let you browse your search term history so that you can combine prior searches into one final master search. For example, you might experiment with eight or nine different variations on a list of terms in the search box before you decide which ones give the best results (i.e. the best balance between **sensitivity** and **specificity**). If you are able to see your search history, along with the number of results each search generated, you will be in a better position to choose the final combination of words and criteria to use. For a worked example of this kind of functionality, see below.

Discuss which databases are most relevant with your supervisor. As with any supervisory meeting, it's probably a good idea to show your own initiative by researching the available topic on your own first, perhaps with the help of your library resources as we mentioned above. You can then ask your supervisor what they think of your plan, rather than simply asking them what they would do if they were in your position.

Constructing a rigorous search

We mentioned above that it is important to think about **sensitivity** and **specificity** with any search you carry out.

We've created a hypothetical search scenario to show you how this works in practice.

Keywords:

efficiency photovoltaic polymer

Search

Results: 17,094

The default way a search engine will interpret your search string is to assume you want to find articles that contain *all* of these words.

This is far too many hits for you to be satisfied with your search (see the section on 'How many results should you aim to get?' on p. 48), but before we can discuss what to do with such a large result, we need to look at the concept of **Boolean operators.**

Boolean operators

In logic and in computing science, these are the extra commands that determine whether your words are interpreted separately or together.

> ### The three classical Boolean operators
>
> - AND
> - OR
> - NOT

If you don't use any at all, the algorithm will usually assume you want to include an **AND** between each term. Results of multi-word searches like the one on the previous page will be more specific than the combined results of searching for each of those three terms separately:

Individual search term	Number of results
efficiency	5,417,349
photovoltaic	269,989
polymer	2,067,012
Combined results:	7,754,350

This doesn't make either strategy inherently better or worse than the other, as that depends on your purpose. If you are already certain you want to find articles where *all* of these three terms are included, then go with the first search approach (to include an **AND** between each word, or to leave nothing between the words, in which case the algorithm will assume this on your behalf). If you want to experiment with the best word to use in each position, however, stick to separate searches for each word and combine them later.

Why would you need to experiment with the best word to use?

'Photovoltaic' might be a term you use in your research project, but there are many other words researchers could use for the same concept. To capture more of these, you can use the **OR** operator.

Keywords:

photovoltaic **OR** photoelectric **OR** optoelectronic

Search

Results: 561,522

The **OR** operator means that articles with *any* of these three terms will produce a result. As you can see, the number of results using this approach is greater than the search for 'photovoltaic' alone (269,989 hits – so this is now more than double), and so we can say the search has become more **sensitive**.

The **NOT** operator can be used in a similar way to exclude results that contain specific words.

Keywords:

polymer **NOT** "plastic products"

Search

Results: 1,781,597

As you saw before, searching for 'polymer' on its own returned 2,067,012 results, so this NOT function has reduced the number of irrelevant results that mention polymers in a fairly common context that we're not actually interested in. (Note that the search algorithm can't distinguish whether or not the word 'polymer' is used in an article to *mean* a plastic product; it simply rejects any results where the phrase 'plastic product' is *also mentioned*.)

Note the position of quotation marks here. This is crucial when you want the search engine to interpret the words as part of a complete phrase. As we showed you above, if you simply don't put any operators between your terms then the algorithm is likely to interpret them as having an **AND** between them, which could create unintended results. Depending on the database you're using, you might need to use single quotation marks, double quotation marks or brackets. Check for specific instructions on your database's help pages.

Combining separate terms

To revisit our initial example, we might now have decided that the best terms to use are:

> polymer **NOT** "plastic products"
> photovoltaic **OR** photoelectric **OR** optoelectronic efficiency

If the database allows you to enter these as separate search strings which it will combine into one search, then you can do so. If not, you need to enter them all in one box and tell the search engine where one search term ends and the next begins. Just as in mathematical expressions, you can do this with brackets:

> (polymer **NOT** "plastic products") **AND** (photovoltaic **OR** photoelectric **OR** optoelectronic) **AND** efficiency

So, in summary, here are the various search strings we've mentioned, and the number of results they return on our database of choice, at the time of writing. We've ordered them in a way that this time might represent the developing thought process of a researcher approaching the search phase by trial and error, as is normal.

(Note that **none of these are what we would regard as a finished search strategy**, as all of the results are still **too numerous** for you to scan through. For more on this, see the section 'How many results should you aim to get?' on p. 48.)

Individual search term	Number of results	Comments
Efficiency	5,417,349	Initial searches, allowing you to gauge how many articles exist on each sub-topic
Photovoltaic	269,989	
Polymer	2,067,012	
(Combined results of the three above)	(7,754,350)	(Too many to read)
Polymer **NOT** "plastic products"	1,781,597	Improves the **specificity** of the polymer search term
Photovoltaic **OR** photoelectric **OR** optoelectronic	561,522	Improves the **sensitivity** of the photovoltaic search term
Efficiency photovoltaic polymer (i.e. algorithm assumes an **AND** between each word)	17,094	Our first demonstration; too vague
(polymer **NOT** "plastic products") **AND** (photovoltaic **OR** photoelectric **OR** optoelectronic) **AND** efficiency	21,088	Our last example; much more robust

Bonus search tools

You'll probably find that your database allows you to use a few extra operators, but the specific usage rules surrounding these tend to vary from database to database. Check the instructions for each specific one to find out precisely how to use each function.

> *Truncation/wildcards:* Useful for finding any words that start/end/contain a string of letters.

If you want to find articles that mention either synthesis, synthetic, synthesise, and the spelling variant synthesise, you can usually shorten the word to the common letters and use a special character to represent any other combination of letters. In this case, the three letters *syn* are present in every variant. If the asterisk symbol was the wildcard for your database, typing *syn** would return articles that include all of the words mentioned above.

However, it would also include any other words that start the same way (synergy, synesthesia, syndicate, etc.), so be careful how liberally you use this option. A search string such as *synthe** would be a better example of good truncation here.

> **Proximity:** *Useful for finding pairs of words that are close to each other but not directly adjacent.*

The final version of the search we showed you in the worked example above would return lots of irrelevant articles about completely different topics as long as they mention both of the words 'photovoltaic' and 'efficiency' anywhere at all.

To reduce the number of these spurious search results, some databases allow you to specify that your search terms must appear within a certain distance from each other. For example, *photovoltaic* **NEAR(9)** *efficiency* would only return results where these two words were no more than nine words apart. Thus, phrases such as '*photovoltaic component efficiency*', '*the efficiency of such photovoltaic components*' and '*efficiency is an important consideration when selecting a photovoltaic material*' would all trigger hits. Articles where the words 'photovoltaic' and 'efficiency' were always separated by more than nine words, however, would not.

The precise characters to use for truncation and proximity searches will vary from database to database, for example **NEAR/9** or **NEAR[9]**, so check the guidelines for yours carefully.

How many results should you aim to get?

We've become accustomed to general internet search results that number in the millions. Out of those millions, how many results do you actually look at before you decide which page to go to?

This is fine for everyday searching because we rely on the algorithms to prioritise the most relevant results first. **Academic databases don't work this way**. Their default is usually to return the most recent results first. If you're used to scanning the first two or three pages of results before moving on, you're likely just finding articles that are about to be printed, and those that have come from the last few months. There's absolutely no reason to think that these are the only relevant articles.

In academic searching, **you should aim to generate a small enough number of results that you can realistically scan through all the titles**, along with the abstracts of the more promising ones, and mark some of them for a proper follow-up.

How many titles could you evaluate in half an hour? Thirty minutes of reading several hundred article titles might sound like a lot of repetitive work, but the stakes are high. Success in your PhD is based partly on having a solid base of literature, so you should make the time to dedicate to searching like this.

Documenting your search strategy

For PhD theses in some disciplines – particularly those where an analysis of literature *is* the research – it can be very important to record how you collected your sample of sources.

Traditional laboratory research is documented in a methodology that makes it possible for future researchers to replicate an investigation. Documenting a search strategy is the equivalent when the research is based mainly on the published findings of others. This type of research includes:

- extended literature reviews/library research projects
- systematic reviews
- meta-analyses.

Each of these is essentially an extra layer of scientific analysis laid on top of all of the individual journal articles published to date on a particular topic. It won't be possible for an author to review every single article in their field, so they'll choose criteria to narrow the scope of their investigation. Choosing these criteria forms part of their methods.

Exclusion criteria

They might decide on their search terms as described above, and then choose to eliminate some articles from the pool of results because those studies have relatively small sample sizes. They might choose to eliminate articles based on a particular treatment protocol, for example, if a drug was administered and the effects were monitored for varying lengths of times in different articles, the reviewer might exclude any that measured the effects for less than six hours, or that did not control their experimental design for other drugs that they suspect could interfere with its action.

A helpful example of a specialised principle

There are a handful of widely known strategies that make good sense for ensuring a search is sensible, thorough and focused. One of these is known by the acronym PICO:

PICO (Problem/Intervention/Comparison/Outcome)

It's a checklist to help you think about the different elements of a study you might want to search for when constructing your search terms, and it works well for any research where you're trying to investigate a cause-and-effect relationship. (It generalises nicely across the sciences, but if you're in medicine, where PICO originates, note that the P can also stand for Patient or Population, and the I can also stand for Indicator.)

> ### Examples of research questions
>
> *Do six-monthly check-ups improve five-year survival rates for prostate cancer in men in Brazil?*
>
> *Does over-activation of small heat shock proteins improve heat tolerance in staple crops, such as rice?*
>
> *Does the use of an ultrafiltration step in bone collagen pre-treatment improve the accuracy of stable carbon and nitrogen isotope measurements?*

The research question would first be broken down into its four PICO elements as follows:

	P (problem/ patient/ population)	I (intervention)	C (comparison)	O (outcome)
Medicine	Men in Brazil	Six-monthly check-ups	Annual check-ups/no routine check-ups	Prostate cancer – improved survival rates
Biology	Rice	Modified hyperactive heat shock proteins	Wild type heat shock proteins	Improved thermotolerance
Chemistry	Bone collagen	Ultrafiltration	Fibreglass filtration	Removal of contaminants/ improved accuracy

You would then list all the search terms you think would apply to each element. For example, 'improved thermotolerance', from the biology example, might be covered adequately by:

thermotolerant OR thermotolerance
'heat tolerance' **OR** 'tolerant of heat' **OR** 'high temperature tolerance' **OR** 'tolerance of high temperature' **OR** 'tolerance of high temperatures' **OR** 'heat stress' **OR** 'thermal stress' **OR** 'high temperature stress'

You can see that this list is quite long, and we certainly haven't included every term that might signify the same concept. Don't worry about including too many search terms, though; the search algorithm doesn't care how many you use. (Note: we haven't used truncation in this example, because we want to emphasise just how widely you should think about casting your net when you create a good academic search. If you missed our explanation of truncation section, you'll find it above.) Adding more terms will make your search more **sensitive**. Just be mindful of balancing this against **specificity**. Sensitivity is achieved in part by combining all of these terms, which only represent the O in PICO for this particular scenario, with the rest of them in the following format:

> (all your 'P' terms) AND (all your 'I' terms) AND (all your 'C' terms) AND (all your 'O' terms)

As we mentioned above, your aim should be to bring the number of article titles returned to a number that you can realistically read. **You could probably scan through 200–300 article titles in half an hour**, deciding as you go whether to mark them for follow-up.

Searching beyond the literature review

Regular searching is a useful habit to develop. Perhaps once every month or two, spend half an hour at your computer checking to see if anything new has been published in your field. This is basic good practice because it might give you new ideas about how to develop your research, but it'll also help soften the effect of every PhD student's worst nightmare: **being scooped.**

This is when someone else publishes an article dealing with the same research question as you. The impact can range from a minor emotional inconvenience, because you'll arrive at your viva and you'll be presenting something that isn't an entirely novel contribution to the field anymore (though this won't necessarily be regarded as your fault), through to undermining your ability to publish a paper, because it's already been done.

If someone scoops you at the end of your PhD, your examiners will understand that there's not much anyone could have done. Your assessment won't be affected by it and you'll still be evaluated on the basis of your progress as a trainee in your scientific field.

If, on the other hand, someone publishes an article on your topic one month after you complete your literature review, it'll be significantly more embarrassing and your examiners might doubt your credibility as a researcher.

It certainly isn't the end of the world (or your academic career), but it's obviously something to try to pre-empt by checking your academic databases regularly to stay abreast of new developments.

Automated search alerts

There are a variety of automated ways you can stay up to date. Major journals will often let you sign up to a mailing list, providing you with a table of contents for each new issue. Some databases will let you save a search configuration, enter your email address and email you with new results at regular intervals. As with all mailing lists, though, human nature is often to de-prioritise opening these messages. We've just described how the stakes with this are high, though, so if you set up automated alerts, schedule some regular time to read them.

Insights from a researcher

My supervisor started me off with a library of papers by sharing it through Endnote. I then used Pubcrawler to get weekly updates on publications relevant to my research. Attending conferences also helped me to identify groups to follow and revealed competitors and collaborators.

MeSH headings (medicine and biomedical sciences)

There are almost always several different words for the same concept. Each scientific field then comes with its own set of terminology for the substances, procedures and theories that exist within it. In the biomedical sciences, however, there is also a hugely expanded vocabulary to describe drugs, treatments, anatomy, biological processes and demographics. This vocabulary is often derived from a combination of Greek and Latin prefixes and suffixes, all of which makes it even more difficult to anticipate all of the appropriate synonyms to use in a database search.

To help address this difficulty, the curators of PubMed (one of the most popular medical databases, short for Public MEDLINE) have created a system of categorising keywords into a master database containing all of the synonyms you might want to use. This system is known as **MeSH – Medical Subject Headings**.

If you enter your first, most obvious search term and then ask the database to map this against the MeSH headings, you'll be able to include all of these synonyms at once. The system is organised hierarchically in a tree structure, meaning your chosen search term might have lots of sub-headings within that area of research, and those might also have sub-sub-headings, too. If you wish, you can fine-tune the search to exclude a specific branch of the hierarchy to increase the specificity of your search.

This means you get to save lots of time that you would otherwise have spent thinking of search lists. Since the terminology for various medical concepts can vary from country to country (e.g. drug names), it also means you can worry a little less about missing out an important synonym you were unaware of.

Keeping abreast of new literature

You should think about returning to carry out your most useful or productive searches again later in your PhD. Your literature review is probably one of the first tasks you'll be asked to do, but if you never return to find new articles until it's time to finish in the lab and focus solely on writing up, you'll have missed a few years' worth of new developments. These could have been useful (or even vital) for the success of your research, and your prospects of getting any publications out of your work depend partly on spotting when someone else publishes something very similar to what you plan to publish.

Hopefully, this chapter will have given you enough advice that you can create good searches, save the best ones and return to re-run them at several points throughout your PhD.

Insights from a researcher

If I could go back in time and talk to myself during the first year of my PhD I would tell her to sign up for literature alerts, and to write a few sentences summarising every relevant paper she read and to add it to a referencing tool that she was familiar with (like EndNote) as she went along.

Insights from a researcher

For me it was definitely organic. I knew a lot of the literature anyway because I had been to conferences or knew the authors. I had also previously done a master's in a related area so that helped. Twitter is also a great source of papers. As part of our doctoral training, we did get taught how to use library databases, which was useful. The main problem is when your institution doesn't have access to the papers you need. That slows things down a bit.

CHAPTER 5 Organising and Keeping Track of the Literature

Writing your PhD thesis is probably going to be the largest single exercise in handling numerous sources that you ever carry out (perhaps topped only by the process of writing review articles, if you progress in academia). It's important, then, to make sure you have a system of not just *collecting* reference details, but also of *collating* them.

The difference is that by collating them, you create a system of organising and annotating the entries such that it's easier to find specific ones again later. If you annotate well, it also means you'll be able to come back months or years later to use one in your writing without necessarily having to read the whole article again to remind yourself of its main points. As you'll become aware, it's usual for your lab materials to stay with your lab after you complete the project so that the next person to pick it up can continue where you left off, so a third benefit of good annotation might be that your lab colleagues and supervisor(s) will be able to make better use of your materials after you leave.

We'll describe how reference management software can help you build a library of the sources you've read, organise them and automate the process of getting those references into your written work. The technical elements are one side of the task, but the choices surrounding how to sort and group your references need to be made by you. We'll describe some commonly used strategies that we've seen PhD students use. We'll describe what we can see as the pros and cons of each, and we won't recommend any as the single best way; we want to give you the tools to be able to design your own approach to reference management given the specific nuances and requirements of your project.

Ultimately, the purpose of managing your referencing information is to help you build and weave a well-supported argument through your thesis. Whether you read just our advice on the technical tools or just our advice on the strategies for grouping and categorising your readings, bear in mind that this should always be your overarching goal while you figure out what works for you.

Insights from a researcher

My strategy for dealing with literature varied depending on what I needed. I mainly collected papers as PDFs and used software to annotate them. Broadly, I saved them in folders and subfolders relating to the different sections and subsections of my introduction. These also correlated to the structures of my results chapters. I used a different approach when I was writing about different DNA assembly methods and trying to find examples of research in which these DNA assembly methods had been used. I searched a literature database and skimmed a lot of papers to find answers to specific questions: what did they assemble; and what did they use it for? It would have been a waste of time to save all of these papers as PDFs so I wrote down the references and answers in a notebook by hand. It felt organic at the time, but I had a minor panic when I thought I had lost my notebook. I should have made a record of them in reference managing software at the time.

General principles of collecting reference information

There are different pieces of terminology we could choose to talk about this process. For the sake of clarity, we'll use **collecting** sources to mean the act of *recording* the referencing information that goes with your source materials somewhere. We'll use the term **database** to refer to any place you choose to store the referencing information itself, which could be on paper, in a card index, in a text document, in a spreadsheet, in a table embedded in a file somewhere, in an app or in dedicated database software.

The reasons for referencing in academia will most likely have been covered extensively in your earlier degree(s), so we won't spend long in this section talking about the rationale for doing it. We do want to recap the reasons for including some of the more obscure elements that go into a reference list entry, though. You don't want to accidentally leave any of them out during your collection phase only to realise later that those particular elements were going to be of use to you, at the re-reading and editing stage, or to a reader, who may or may not have access to the material in the same format as you.

Surnames	To indicate authorship, in a pre-defined order indicative of level of contribution to the article
First names/initials only/ middle initials	To differentiate between academic authors with the same surname
Article name	To describe the content of the article
Journal	To provide the first level of information about where to find the article; also carries implicit information about the impact of the article
Volume number	To help readers navigate the back catalogue of a journal, based on year
Issue number	To help readers navigate the back catalogue of a journal, based on the multiple volumes that were created within that year
Page number(s)	To help readers with access to a whole issue quickly navigate to a specific article
Digital Object Identifier (DOI)	To allow quick and simple online access to an article without relying on more conventional online identifiers, such as a web address, which could change and thus would not be futureproof.*

*For this reason, URLs are not normally used in referencing unless your source exists only in web page form. Note that articles downloaded from journal websites do not fit this description. Newer online-only journals are something of a mixed case, but they will usually closely follow the traditional academic referencing format. Use all of the volume and issue information that they supply, just as you would with any other article.

Reading an article today and writing about it in three years is a challenge of organisation and of memory. Future-you will appreciate the efforts of present-you in recording *all* of these referencing details now, as there's no guarantee that shortcuts such as bookmarks to specific web pages will still work that far into the future.

When deciding where to collect and record all this information, think about the way you're likely to want to think of the articles later.

- Do you tend to remember articles based on the name of the lead author?
- Do you care about which articles are the most up to date?
- Do you categorise articles on the basis of the journal they were published in?

- Do you remember articles by their titles?
- Are you more likely to remember them by pieces of their contents, that is, by words and phrases that might not appear anywhere in the referencing information at all?

If you're going to create an index of your articles, we'd strongly suggest that the most useful format would be one that will let you sort, filter and group the entries. A text document is searchable, but it's not easily sortable. A spreadsheet is sortable, but it comes with limitations to how much information you can see at a glance when everything is stored in rows and columns of fixed width. Database software will let you sort and filter in the ways we mentioned above, but it'll also have a much more user-friendly interface than a spreadsheet.

Although you might not think of it as a database, **reference management software** – or simply a **reference manager** – is this kind of solution.

Reference management software

There are many different individual pieces of referencing software on the market. Some are free, some will have plugins for word-processing software, and some are more well-supported by their developers than others. We don't want to recommend a specific company's product, but we do want to outline what features you might want to take into consideration, and how you should use it.

Good practice when using a reference manager

Reference managers exist primarily to store all of your referencing entries for you to access later. Each time you choose to read a source, we recommend that you **store its reference information immediately**, even before you sit down to read it. You might not decide that the article is relevant today, but your project might eventually develop into an area where you realise it becomes useful later. If you have all of the source's details recorded in your reference manager, you'll be in a better position to find it again at that point when all you can vaguely remember is the author's name or one of the words in the title. Searching your reference manager's library will be much more efficient than searching an online database's entire indexed collection.

Getting information about your sources into your reference manager

When you enter details of a source into a reference manager, you won't be able to simply type in the whole reference in the form you'd use at the end of your thesis, for example:

> Singh, V. and Mayer, P. (2014) Scientific writing: Strategies and tools for students and advisors, *Biochemistry and Molecular Biology Education, 42,* 5, pp. 405–413.

Reference managers need each piece of information separately. In database terms, each **field** needs to be entered on its own. This allows you to sort and filter your list of references, and it also means the software can reformat the data to produce a reference list in whichever referencing style you need. For example, if you use the Harvard format while drafting your work and then discover later that your department would prefer you to use Vancouver, a reference manager can take care of this for you – but only because it stores all of the information in separate database fields.

There are various methods to actually get a reference into your manager. You'll be able to type it in directly by creating a new entry in your database (usually called a **library** by the software). Typing the information yourself is the most time-consuming method, but also sometimes the most reliable, as other methods rely on data being transferred from one computer system to another. These automated methods can sometimes introduce errors in interpretation, for example when one system outputs data in a format that doesn't match what the other system expects. When you enter it by yourself, field by field, you can be sure that *all* of the relevant information is captured (as long as you don't make any typos...).

You'll probably also be able to import the referencing information by supplying the reference manager with a file. These files are the ones you'll find as downloads from the webpages of journals. When you've found an article you want to read and you see the 'download PDF' option, look for another one close by called something like 'export citation'. The file you'll get contain all the fields your reference manager needs to create an entry in your library, encoded in a way that easily identifies each separate element/field of the reference.

When you hit that download button, you'll usually then be asked which format you need. Different reference management software companies have created their own standards over the years, and you'll need to find out which format your own reference manager expects you to work with.

To illustrate the differences, here's the referencing information for the article we mentioned above formatted in three of the common formats at the time of writing:

BibTex format

@article{singh2014scientific,
title={Scientific writing: strategies and tools for students and advisors},
author={Singh, Vikash and Mayer, Philipp},
journal={Biochemistry and Molecular Biology Education},
volume={42},
number={5},
pages={405–413},
year={2014},
publisher={Wiley Online Library}
}

EndNote format

%0 Journal Article
%T Scientific writing: strategies and tools for students and advisors
%A Singh, Vikash
%A Mayer, Philipp
%J Biochemistry and Molecular Biology Education
%V 42
%N 5
%P 405–413
%@ 1539–3429
%D 2014
%I Wiley Online Library

> **RefMan format**
>
> TY – JOUR
> T1 – Scientific writing: strategies and tools for students and advisors
> A1 – Singh, Vikash
> A1 – Mayer, Philipp
> JO – Biochemistry and Molecular Biology Education
> VL – 42
> IS – 5
> SP – 405
> EP – 413
> SN – 1539-3429
> Y1 – 2014
> PB – Wiley Online Library

You don't need to know these formats, but you'll see that there are clearly differences that'll cause problems if you download your data in a format that your reference manager isn't designed to understand.

If you're working with sources that aren't academic journals, you probably won't have the luxury of these industry standard citation files. In those cases, you'll probably have to enter the details by hand. Some software will let you do things like enter the ISBN of a book, or scan its barcode to find the publishing information. Some reference managers also have plugins for web browsers that scrape information from webpages, for example an online shopping page where a book is listed.

Whichever method you use to get data *in*, it's important to manually check that the information's complete and correct before you move on.

Checking your reference manager's library is complete and accurate

You don't want to proceed all the way to printing and binding your thesis before discovering that you've only got partial information for some of your several hundred references. Even if you've properly downloaded and imported a citation file into your reference manager, there's no guarantee that the file's information was entered properly and completely by the staff at the journal.

Think about the pieces of information you'll need when you create your reference list:

- Do you have the authors' first names, **or only their initials**?
- Has the journal's name been given in the **abbreviated form**, for example *Int. J. Theor. Phys*, instead of *International Journal of Theoretical Physics*? Will you need to use the full version of the title to satisfy your university's requirements?
- **Is every word in the title capitalised** (*title case*) or is it only the first word (*sentence case*)? Does this match the format you need to use in your thesis?
- Do you have **the *page range*** for the article, or only the number of the first page?

Books and other source types will require other reference elements (other *fields* in the database, or library).

We won't go into the finer details of how to reference every type of source, as these rules will vary based on the referencing style that your institution wants you to use (Harvard, Vancouver, APA, etc.), and we simply can't cover them all.

Further, some of these styles are overseen by professional bodies and are updated every few years to take account of new types of sources (e.g. APA style, which is dictated by the American Psychological Association). Our advice would therefore eventually go out of date. In contrast, some styles are very common yet not actually centrally maintained (e.g. Harvard), and so very many variants on a central theme can exist.

Rather than taking our recommendation, check first with your department to see if they have any rules. If they don't, ask your supervisor which style they would recommend. You'll find a thorough explanation of various different citation styles in the excellent *Cite Them Right: The Essential Referencing Guide* (Pears and Shields, 2019), and the accompanying website, www.citethemrightonline.com.

If your university tells you that the choice is yours, pick whichever you think most closely represents your membership of a particular academic field based on the style's prevalence in the literature you read (or simply pick the one you think is clearest and most useful for your reader!)

Getting references back *out*

As with getting data *into* your database, there are numerous different ways that your specific reference management software could be designed to give you the data back *out* at the end. Rather than trying to give you a step-by-step walk-through, we want to make you aware of the key general benefits they should be able to bring.

Creating a reference list

Your thesis will have several hundred references, which will take several hours of your life to type correctly if you do it by hand. The primary benefit of using a manager is that you don't need to type out every single reference into your thesis by hand.

Keeping the list in order

Adding references as you write will mean finding precisely the right place for them in the growing list, which introduces the possibility of human error. This is especially important if you're using a numbered system, as any late additions will mean changing the number of every reference that follows.

If you've to use a numbered format, we'd argue that this benefit alone is enough to justify the time it takes to learn how to use a reference manager.

Making reference lists from only some entries in your database

If you're diligent, your reference library might end up containing details of every source you read during the course of your PhD. You obviously won't be including all of them, as some won't have been as relevant or as useful as you thought they might be. Rather than deleting details of the articles you choose not to incorporate into your thesis, your manager will let you select a subset of your library when you make a reference list. Alternatively, you might be able to group and label the references in your library by the chapter or topic that they're most relevant to. You can then use these groups or labels to make a formatted reference list from only the subset of your library you want to use.

Reformatting references in a different style

Will you use an alphabetical referencing format (e.g. Harvard or APA), or will you use a numerical one (e.g. Vancouver)? If you start in one format and realise later that you have to use the other, your reference manager can take care of the reformatting for you.

Inserting citations into your thesis as you write

You might find that your reference manager has a plugin for your text editor of choice. These plugins let you hit a button, search for a reference, and incorporate it into your document wherever your cursor is at that moment in time. This hugely simplifies the process of inserting details of sources, as both the in-text citation and the full reference at the end will be managed and maintained by the computer. Again, hopefully you can see that the minute or two (or five) that it might take you to manually incorporate one source by hand quickly multiplies up into several hours of your life over the course of a whole thesis, making automation a worthwhile investment of your time.

Annotating and grouping your references

Once you've got a reference stored in the management software, you'll probably want to start annotating and categorising it. You should be able to file each entry under a different grouping.

You could do this on the basis of scientific topics or themes. If your project has several strands to it, you may plan to have a different chapter for each one. Your reference group themes might then reflect the structure of your PhD.

Alternatively, you could group them on the basis of the techniques they describe. Your methodology section will probably end up with subsections to describe the various different types of work you carried out. You might have a section for lab methodologies and another for fieldwork methodologies (this means you can group individual methods in a logical, sequential order, thus helping your reader follow your processes more easily). Each of the methods (the individual processes) might be adapted from literature you've read. Grouping those literature references in your database on the basis of these two broad methodology sections would make your life easier when browsing your library and deciding which articles you need to use as references in your thesis.

The alternative would be to read through the titles and abstracts in your reference manager one-by-one until you identified all the articles you thought were relevant. This might not be obvious from the titles alone, particularly if the titles have nothing to do with the techniques they describe inside, which might be the only reason you downloaded the article in the first place.

Another method might be to group them on the basis of how reliable you found their contents – though we'd suggest that this shouldn't be a primary way to group them. Reliability is of course important, but we propose that it'd be more useful to use a categorisation that describes the science they contain.

CHAPTER 6 # Reading and Critiquing the Literature

This chapter relates to one of the biggest challenges of doing a PhD: *engaging critically* with the literature. Aside from committing to an independent management of your workload, this is perhaps the biggest step change expected of you from your previous studies, whether that was at master's or undergraduate degree level.

Each of those two words – engaging critically – deserves to be examined separately.

Engaging means not just reporting what's been found in other sources, but working this into the flow of your own academic argument. It's not enough to present a series of pieces of evidence from your sources and expect that your reader will reach the same conclusion as you do; your job as a writer is to show the reader how these pieces relate to each other, and to demonstrate your ability to synthesise these as part of a coherent and novel contribution to the wider academic discussion (i.e. your thesis).

Critically means demonstrating your ability to judge each of these pieces of published evidence in relation to each other. Is one piece of evidence more strong than another? Has something been carried out particularly well? Are there shortcomings in one experiment and, crucially, can you suggest what could have been done to improve it?

Simply collecting and reporting what's been discovered is a job better suited to a junior research assistant. It involves *recognising* that material is relevant to a topic, but it doesn't require any deeper analysis of it. Your responsibility is to show your reader how it informs or gives context to the primary research you carry out for your doctorate.

In this chapter, we'll examine what the concept of critical analysis means. We'll also discuss the features your examiners will be interested in seeing, and we'll give you some practical pointers on how to develop your skill in it.

Reading strategies

Most academics will tell you that they rarely read articles from start to finish. They're also probably in the habit of looking for different *types* of articles depending on which stage of the research process they're at, or how familiar they already are with the topic. This section will give you some brief guidance on how you can make life easier for yourself by being strategic about what you read, when you read it, and how you decide what to do with the knowledge you glean.

Choosing which *type* of articles to read first

If you're at the start of a project, you might want to start by focusing on reading review articles, as these are the publications that can give you an overview of a topic. They'll probably be much longer than any primary research papers (i.e. articles that focus on one specific investigation; 'primary' refers to the level of separation between the author and the events being recorded, with review articles falling into the 'secondary sources' category).

Their increased length can often come with a lower complexity, however, as they exist to highlight key points from complex datasets and condense them into summaries and key statistics, which they then interpret and critique for you as part of their review. Review articles are also a good place to start as they'll provide signposts to more literature, much of which you might not have found by searching an academic database because you didn't yet know what words you should even be searching for.

We find that many of our students were directed by their supervisors to read one or two seminal articles at the start of their project. These are often reviews or primary research papers where someone first reported a key finding/developed a theory/used a new technique in their field.

Becoming confident navigating articles

You've probably read many tens or even hundreds of articles before starting your PhD, so we won't go into lots of specific detail about what you might want to extract from each section. (We covered this in greater detail in our previous book: *Writing for Science Students* (Boyle and Ramsay, 2017).)

What *is* worth expanding upon here is the differences of opinion between undergraduates, postgraduates and academic staff on the most important sections of an article.

Hubbard and Dunbar (2017) asked second year undergraduates, third year undergraduates, PhD students, postdocs and established academics to rank the general sections of scientific articles (abstracts, introductions, methods, results (text), results (figures) and discussions) in terms of their difficulty and their importance.

On the topic of difficulty, abstracts and introductions were consistently regarded as fairly easy. These sections were followed by discussions, which undergraduates rated much less easy than either of the previous two categories, but mature academics rated equivalently. The results sections (both text and figures) and methodology sections were regarded as the most difficult to read, and these responses followed a clear trend towards easiness as the respondents progressed from undergraduate stage to academic stage too.

Interestingly, while difficulty rankings *all* tended towards the easy end of the scale as respondents became more experienced, the data on which section was regarded as the *most important* showed that some sections' scores went up while others went down.

Introductions and discussions (i.e. the background and the analysis of an article) were rated as *most* important by undergraduates, but these followed steep drop-offs to the two *least* important by the time respondents were academic staff. The methods section was ranked as the least important section by undergraduates, but this climbed steadily through PhD and postdoc phases to a final rank of third most important section by academics.

We hope this shows that it's normal for you to shift your focus as you gain more experience in reading literature. Once you've established that you're confident in navigating your way around a paper and annotating it as you go, you might want to start being selective in which parts you read first.

Choosing which *section* of an article to read first

You'll read differently depending on your purposes. For example, if you're trying to repeat someone's experiments, you might read in a very different way than if you were reading to gather new knowledge about a new topic.

- **Purpose: to learn about a methodology or protocol**
  - Start by reading the **methodology**. Perhaps you wish to read the results section to check how well the set-up worked, or to find out what format of output you should expect. If you only need to know how to carry out that procedure, you may choose not to read the rest of the article at all.

- **Purpose: to get background knowledge**
  - For most of your undergraduate studies, the bulk of the information you were taught could probably have been found described in a textbook. Now that you're working in a rapidly developing field of research, knowledge develops and changes too quickly for it to all reach textbooks, and so you need to find it in scientific articles. The **introduction** section of an article is the next best thing to a textbook in this case, as it summarises what we've already learned up to the point these researchers started to do their experiments.

- **Purpose: as a source of articles for further reading**
  - Citations will be present throughout all of an article, but the **introduction** is especially rich. For the same reasons as above, an introduction will have many signposts to other previous articles. This can help augment the searching you do on your favourite academic database as there will be times when you're unconsciously unaware of what else you *should* be searching for.

- **Purpose: in-depth research on a specific primary article**
  - In this case, you're probably trying to design your own investigation. You might want to know what others have done in your field so that you can plan a project to pick up where they left off. In that case, start with either the **results** or the **conclusion** so that you know the paper has something important to say. If it doesn't, you can either discard it totally, or you can note the absence of any interesting result and make sure to record the article information in your reference manager so that you can report on it in your thesis later. If the article *does* have something interesting to claim, then you can move on to familiarising yourself with the rest of the sections and critically analysing it in full.

What is critical analysis?

When you undertake a literature review, **you take on the role of critic**. You'll find that we deliberately don't use the word 'criticism' in this book to describe what you should be doing, as that would usually be read as a solely negative description of the task. Rather than *criticism*, think of film *critics*: they are employed principally to rate films from the terrible to the excellent, and usually to give them a sort of score out of five or ten. If their critiquing involved only pointing out the flaws in a film, the critic's traditional five-star scale would only go up to 2.5. In a similar way, **you need to be prepared to acknowledge the positive aspects of scientific literature, too**.

Of course, film critics don't only rank a film on a scale. They give a thorough justification for their final opinion by taking into account the many different aspects that go into creating a good film, and a good critic's review will involve several hundred words of writing to convince you of that justification. Your justifications probably won't be several hundred words long for each article you mention in your thesis, **but you should find yourself writing extensive notes – at least on the most interesting, relevant or controversial papers – during the reading phase of your literature review**.

Broadly, you should be evaluating:

- the ways in which a piece of research was carried out
- the research question the authors were trying to answer
- whether the conclusion addresses that question
- whether you've seen any conflicting information (either in other articles or in your fundamental understanding of that scientific area)
- whether you think the authors have overstated their impact in their final conclusions or discussion.

Again, critical analysis involves noting when and how these were done well. Even if you don't find yourself writing very much about this in the final version of your thesis, you should develop your skill in thinking about it, as it is crucial to becoming a successful scientist yourself.

Taking the film critic analogy to its logical conclusion, the final (perhaps more implicit) role of a film critic is to persuade someone why they should or shouldn't spend their time and money watching a particular film. **The equivalent for you, as a scientific writer and PhD candidate, is to implicitly persuade your reader why they should or shouldn't trust in the various conclusions of an article**. The stakes here can be high. For a scientific reader of your critique, the equivalent of going to see the film could be anything ranging from simply incorporating the conclusions of that article to their sum of understanding about the field, right up to applying for a multi-million-dollar grant so that their own lab group can invest four or five years taking that research to the next step. (Much more expensive than a cinema ticket.)

Why is critical analysis important?

We've just given quite an extreme example, but there are many more subtle reasons why scientists value good critical analysis as strongly as they do.

First and foremost, it helps to demonstrate your original contribution to your field. This is the requirement that underpins PhDs across all subjects: a good research project supported by a well-argued critique of literature is a more complete demonstration of a student's original contribution than a good research project supported simply by a very descriptive report about the related articles. Your examiners will most likely ask you questions about your literature review during your final exam (depending on where you choose to study, this could be called your 'viva' or your 'thesis defence'). This questioning means you'll need to be fully prepared with your own thoughts about the strengths and weaknesses of all the articles you write about.

It's also important that you learn for yourself how to discriminate between what's right and what's wrong, or what's somewhere in between. Science (the scientists, the experimental protocols, an article's peer reviewers, a journal's editors …) isn't a perfect discipline. You can see this in the fact that articles are retracted every month. Even with the majority of articles that remain published, there are pilot studies with small sample sizes and protocols that aren't yet perfect, there are unconscious biases that slip by unnoticed, and there are many other ways a piece of research could be imperfect. One of the defining features of the scientific disciplines is that they are self-correcting; when new information supersedes what was 'known' before, then we update our understanding as appropriate. In becoming a member of this academic community, you take on responsibility for contributing to the self-monitoring and self-correcting that happens across the field. It's not simply a noble endeavour, though; good critical analysis has real-world ramifications.

Fundamental science research isn't done for its own sake. Yes, individual researchers are drawn to projects they think are inherently interesting and intellectually satisfying, but at the level of funding and policy, projects are approved on their likelihood to generate returns on investment. We might not know precisely what those returns are likely to be yet, but, as a society, we investigate things on the premise that there will be a pay-off *at some point*. It's important, then, that we are critical and endeavour to understand what's truly correct so that future research builds upon a solid foundation.

Some science has strong, direct connections to the field of industry. It goes without saying that research done in an industrial, commercial company is only going to be done if it promises a direct financial return, but even fundamental science research carried out by students at universities informs the development of new products, new technologies and new solutions to problems. The self-correcting mechanism of critical analysis therefore plays an important role in making sure these developments can be made at all.

Some scientific research also has an important and direct role in safety. Materials development, drug design, manufacturing and fabrication projects, structural engineering, electronics – these are a few of the fields where academic research is active and is aimed at improving safety for the people touched by each of those industries. Being critical and observant of other scientists' findings helps to make sure they are as safe as they can be.

The field of healthcare is another where the safety implications of uncritical researchers are obvious. Of all the sciences, research into medical treatments and into the allied health professions perhaps has the most immediate impact on safety, as the stakes are so often literally a case of life or death, and because medical professionals are approached with the expectation that they'll be able to *improve* outcomes for a member of the public, not the opposite. Reading medical literature with a critical eye will help to safeguard the prospects for patients in the future.

Hopefully this convinces you to slow down as you read articles for your literature review and pay close attention to whether you agree or disagree. Next, we want to show you *how* to do it.

How can a student like you be qualified to critique an expert's published work?

This is one of the most common questions we hear from our students. They frequently ask how they, at this early stage in their work, can be expected to judge the work of a far more experienced and qualified scientist.

Firstly, **you *do* have your own experience**, and where this overlaps directly with the science you're critiquing, it's perfectly valid to draw on that to judge that science. It may not be a large and varied base of experience (yet), but it exists. If you're not totally sure whether someone has perhaps just published a valid variation on a protocol you're familiar with, ask others in your lab, or your supervisor, or colleagues at other institutions whether they see the same shortcomings as you. Perhaps they'll explain why there's a discrepancy or perhaps they'll validate your critique.

1. **Rely on the fundamentals.** There are basic pieces of knowledge you've been taught in your previous degree(s). These fundamentals – usually referred to as 'first principles' – are usually the rules, laws, equations, natural processes and other undeniable building blocks that lie at the heart of your subject. When you find a piece of research that seems to contradict this established knowledge – these first principles – take it as a cue to closely examine the piece of work and determine whether you can spot any weaknesses that would explain it.
2. **Be sure to consider the raw data.** Scientific articles are written in the way they are so that readers, like you, can consider the raw data before being told what the authors think that it means. In other words, the results section is separate from the conclusion and discussion sections. This has been one of the cornerstones of the standard introduction, methods, results and discussion (IMRaD) format since at least the 1960s. If you were simultaneously presented with a finding and the interpretation, it would be much more difficult for you, as a fairly inexperienced reader, to avoid becoming primed to accept their conclusion.
3. **Come to your own interpretation.** If you're in the habit of reading articles in one go without stopping to think about the answer to some

critical questions, consider pausing to come to your own interpretation of the results before you read the author's. For example, look at the statistical analyses in the article. Do they show a strong effect, or are they close to (or even just below) the threshold of statistical significance? If the latter, don't be swayed by the author claiming a stronger effect in their written discussion than you've observed to be the case in the data.
4. **Ask yourself whether the work directly contradicts other literature.** Regardless of how much you might understand of the details, the very fact that two publications arrive at opposing conclusions automatically means that there is still work to be done in that area. It may not be the case that someone has done something *wrong*; it may simply be that differences in the experimental approaches have caused different results, even when they were both designed to answer the same research question. In this case, try to identify or hypothesise on the cause of the difference. Even if you don't have a solution to the issue, your examiner will appreciate the fact that you've recognised it and thought about what it means.
5. **Remember that there will always be diversity of opinion and interpretation between researchers.** We won't be the first to try to convince you that it would be foolish to assume every discovery made so far represents the absolute and final truth on a particular issue. Science is self-correcting, in that further developments have the power to supersede and overwrite previously known 'truths'. The nature of that continual progression means that many of the findings you read about in the literature will be incomplete, just plain erroneous, or only true in a very particular set of circumstances. We know this doesn't necessarily give you the power to identify specific shortcomings, but it should help you become a little more confident in your right to critique someone's published work.

On what basis can you critique someone's work?

We want to give you a list of questions to think about when you're reading a research article. We mentioned this briefly above, in our third point about why you're qualified to critique the work of others, and here we'll expand on some common ways to find opportunities to demonstrate your critical ability.

Sample size and replication

We wish there were an easy answer to the question 'How many replicates is enough?', but the answer will be different for each experiment. It's generally (and lazily) accepted that one replicate is not enough, that two leaves open the possibility that a discrepancy (or a match) between the results is simply down to chance, and that three replicates allows room for two 'correct' answers plus one erroneous one. However, this is just the same type of assumption we use in games of chance. A best-of-three approach involves an odd number, so there's an inbuilt mechanism to get one 'winner'. If you were to toss a coin to answer the research question 'Which side lands face-up more often?', you'd be able to come to a clear answer because it's not possible to get a 50/50 answer with three replicates; one side has to come up more frequently than the other. Because of the small number of repeats, it also means you have a large relative difference: a result of one head and two tails could be interpreted to mean that tails is twice as likely to come up as heads. This is the same as saying that tails will come up two-thirds (or 66%) of the time and heads only one-third (or 33%) of the time. These numbers can be further restated – validly – that the frequency of a tails result is 200% of the frequency of the coin showing heads. To the naïve reader who has never had a chance to toss a coin (and you should of course substitute your own experimental result here), this gives the impression of a clear-cut difference.

Hopefully you can see, though, that this is hardly scientific and rigorous. It is a symptom of the common fallacy known as 'the law of small numbers'. Why, then, are we so tempted to take this approach to replication in our experiments?

We'll consider sample size and replication. These are two separate issues, and they deserve two different explanations before we begin:

> *Replication* refers to the number of times the process is repeated.
>
> The *sample size* is the number of individual specimens that go through a procedure at the same time.

If an experiment is designed to investigate a cause-and-effect relationship (e.g. adding chemical X to sample Y to see what happens, or heating sample Y to a certain temperature), a rigorous experimental protocol should involve an element of repetition. If the individual samples in the repeats are theoretically identical (see below), the process would be called *replication*. If the samples are inherently variable (see below), the process of repeating the experiment would probably be thought of as *increasing the sample size*.

Examples of **identical samples** might be:

- chemical solutions prepared according to an established standard protocol
- small sub-samples of a mixture all drawn from one master mixture
- repeat observations made on one physical item/phenomenon, for example brightness measurements made on a star.

Samples used for this type of repetition would be termed *replicates*, and their function would be to ensure that the *measurement* and other technical steps are reliable (e.g. a piece of lab equipment that produces different results on identical samples should give you cause for concern, and this would be a cue to investigate the workings of your equipment or your protocol. Conversely, consistent results prove to your readers that your technical processes were reliable and that they can trust your measurements).

Examples of **variable samples** might be:

- multiple offspring from one set of parents
- multiple individual organisms *without* any family connection
- samples of the same thing collected from different locations.

Samples used for this type of repetition would be referred to simply as *different/unique samples*, thus increasing the *sample size*, and the function would be to test whether the different samples respond to the experimental procedure in the same way (e.g. a large sample size is required in a medical trial to ensure that a drug is likely to cause the same positive outcome for as many different – variable – people as possible).

The lines between these two classifications can become blurred. If a new construction material is being developed and the research involved testing its safety and suitability for real-world use, then part of the implicit assumption being tested is that any one of the batches of the new material may be faulty. In this case, multiple samples drawn from one master mix of raw materials and created by the same machinery could in fact be treated as intrinsically variable samples, and you might regard the repeat measurements as each contributing to the sample size, rather than to the number of replicates. In such a situation, replication might come from the number of times a stress test was carried out. Such a design would ensure that both (a) any variability of the batches of the new material and (b) any variability in the rigour of the stress test were accounted for.

Whether you're considering an appropriate *sample size* for an experiment or the optimal number of *replicates*, the question of 'how many is enough?' depends on what's being measured.

If there are many ways in which two samples could differ from each other, it would take a greater number of replicates to find a truly representative average result amongst the noise. For example, the well-established connection between smoking and lung cancer would never have been discovered in an observation of three smokers and three non-smokers. People can vary in an almost infinite number of ways, and the number of reasons why a person may develop lung cancer or not is huge. Large-scale observations on many *thousands* of people were necessary for researchers to be able to draw reliable conclusions about that particular relationship.

For more specific advice on sample sizes, we recommend investigating the concept of *statistical power*. Broadly speaking, statistical power calculations indicate whether a sample size is large enough for reliable conclusions to be made. Statistical analysis is beyond the scope of this book, but investing the time to learn about statistical power will help you interpret the findings of others, as well as to design your own experiments.

Study design's relationship to the research question

If someone sets out to answer a question, their ability to do so hinges on their use of an appropriate methodology. Some methods just don't have the power to find the answers to certain questions. In medicine, for example, a 'cross-sectional' study is one that establishes the situation on a large population at a specific point in time. A survey might be posted out to every resident living within a local authority's borders, with questions about their smoking, eating and drinking habits, and about their cardiac health. At first glance, you can see that you would be able to draw conclusions about the connection between certain lifestyle choices and health impacts. However, it might not give an accurate report of how long these people have been in those habits. It also might not tell you how long their health situation had been the way it currently is. Since the nature of these lifestyle choices is that they take many years to cause an effect on health, a cross-sectional study is only of limited use, and the questions would need to be carefully worded so as to ask about histories over the past five or ten years in a way that people would genuinely be able to remember the answers. Much more sensible would be a 'longitudinal' study design, which follows people for a much longer period of time and asks participants for their responses to the questions at regular intervals. Cause-and-effect relationships would be much easier to tease out in this situation.

When considering a study design's appropriateness, ask yourself:

- Was the study carried out for a long enough period?
- Was the most appropriate equipment used for the question being asked?
- Was the most appropriate procedure carried out to prepare and process the samples?
- Were measurements taken at an appropriate frequency to capture the fluctuations under study?

Use of controls

The usual list of controls consists, of course, of 'positive' and 'negative'. Before asking yourself whether they are missing from the articles you read, take a second to remind yourself exactly why they're useful – not every experiment needs (or *can* have) them.

Positive controls are important where an experiment might plausibly cause no effect, or where someone might plausibly detect no output because the effectiveness of the procedure is not yet known (we'll refer to these various detectable outputs collectively as 'signals'). In those cases, one must be able to prove that the absence of a signal is not because of a failure in the procedure – a **false negative**. If the positive control returns a signal and the experimental sample does not, the experimenter can say with confidence that this was a genuine finding.

Negative controls are important where a signal might plausibly be detected regardless of whether the experimentation caused an effect.

If the experimental set-up might be contaminated by something, or if a detector is in some way overly sensitive, then a negative control would reveal these findings to be **false positives**.

So, what kind of experiment requires a positive control, and what kind requires a negative? Or, rather, what kind *doesn't* need one or the other?

Controls are useful at the level of individual protocols, and they show that individual steps have been applied properly. At a slightly higher conceptual level, though, it can sometimes be difficult to find equivalent safeguards. For example, in a drug design trial where a totally new chemical synthesis has created a novel molecule, what would you use as a positive control? If the aim is to create a molecule that carries out a function in the body that we currently have no drugs for, a failure would simply have to be regarded as a genuine negative result. The researchers might try again with a range of different chemical conditions and a range of variations on the molecule, but the reader of any write-up would need to satisfy themselves with *indirect* controls, that is, those from individual parts of the overall process (lab measurements showing that the chemical was produced in satisfactory quantities and at satisfactory purity levels, and that it was shown to interact with its target molecule as expected in a test tube situation, all before it was attempted in a whole organism).

There are other situations where the need for specific controls isn't as clearly defined as you might hope. As a critical reader, you should think about these situations too. Rather than calling them 'controls', we'll refer to these as …

Use of 'optimisations'

Moving from a bachelor's or master's degree into a prolonged piece of independent work in an authentic research setting, many students are struck by how much of scientific practice appears to be made up as the researchers go along. This isn't evident from any of the articles you'd read, as we're all trained to write up our work in a very matter-of-fact tone. This encourages researchers to use euphemisms such as 'samples were incubated at 4°C overnight' to mean 'samples were put in the fridge because it was the end of the day and I knew it would be okay to pick up where I left off the next morning'. There's nothing wrong with this! But it does mean that a great many scientific protocols are built upon improvisations and steps timed for convenience rather than for strictly scientific reasons.

On a more serious angle, if a researcher has created a completely novel protocol to suit their specific set of variables, they may simply present the final experimental design they came up with as though this was always going to be the best way to do it. For example, if something was left running for a certain length of time, do the authors explain why they used that specific duration? Or, perhaps more importantly, do they say what range of durations they attempted, and thus why they settled on this final one?

These types of steps are called 'optimisations', where a procedure is refined to give the best result possible. Strictly speaking, they work differently to the positive and negative controls we mentioned above, but they do serve a similar purpose. They help to delineate the boundary conditions where an investigation would be expected to work and where it would not.

When you read articles, ask yourself whether you understand why certain conditions have been used. Have the researchers mentioned the outcome of any such optimisation process? If not, why do you think this might be? Can you see a reason why a different set-up might have yielded even better results? There may be a genuine reason, so don't assume that the absence of any optimisation details means a paper is lacking. However, it would represent an unanswered question, which you can add to your list of discussion points when you write any critique of such an article. (Indeed, there's nothing to stop you from emailing to make enquiries by email – just make sure to approach the scientist in a fashion that you'd wish to be approached by a reader of your own work. If you're motivated enough to do this, it's probably because you're thinking of repeating or building upon the experiment in your own work. In that case, let them know this so that they can see you are approaching them in the spirit of scientific collaboration.)

Recency of cited material

More up-to-date research doesn't automatically mean *better* research, but if you find that an article relies heavily on old sources, it might be a sign that the authors didn't build their work upon all of the existing knowledge available to them. Finding the most current research seems like an obvious thing to do, so failing to lay that groundwork suggests that either the authors aren't as rigorous as they should be, or that they're deliberately trying to conceal something.

Authority of the author

Journal articles (as opposed to websites, etc.) are almost always published by qualified scientists or higher degree students, and so the question of someone's authority to speak as an expert on a topic almost always has to be taken for granted. If you use material from outside of traditional academic literature, though, be careful to assess the credentials of the author(s). If you can't find out what qualifies them to speak as an objective, evidence-based source, then you shouldn't place as much faith in what they say as you do with peer-reviewed journal article authors. This also applies to textbooks: anyone can publish a book if they're willing to pay for it to be done. Websites and ebooks can even be published at zero cost.

Don't mistake someone's lived experience of a situation for academic, qualification-backed experience in recording and critically evaluating the facts around that situation. Websites that claim to represent organisations and other such bodies (e.g. political lobbying groups, disease support networks and charities) are not automatically on the same level of trustworthiness as a scientist who researches the topic. These groups may present evidence, but the very fact that they exist to further an agenda means bias is highly likely to creep into their output. (Be particularly wary of any website that has items or services available for sale!)

Pressures to publish bad science/why bad science gets published

There are many pressures that drive authors to publish their work (perceived obligations to funding bodies; high expectations from universities hosting the labs; career progression, etc.). This means that practices such as cherry-picking of data that supports a definitive conclusion, deletion and adjustment of data that does not, and image manipulation take place.

In addition to the authors who are keen to publish low-quality or misleading science, there are journals that are keen to receive them. Low impact-factor journals receive fewer submissions than those with higher impact factors, and so an author *might* find the selection process to be less rigorous in a smaller journal. They may find it easier to get unconvincing data past an editorial board who wish to publish articles and thereby increase their journal's circulation, and being a smaller journal, they may not have the prestige to be able to convince the most qualified academics to serve as peer reviewers. This means the articles submitted to them may not receive as rigorous an appraisal as those submitted to leading journals in their field, thus increasing the likelihood that 'bad science' will be printed.

In most cases, you won't be able to detect these dishonest practices as you don't have the facilities to repeat the experiments. It's important that you know they take place, though, and that they are entirely to be eradicated wherever you find them so that science continues to be based upon empirical truth, and to ensure that it retains the public's trust.

One resource you may wish to consult is Retraction Watch, a project backed by the Center for Scientific Integrity aimed at highlighting retracted papers from across the scientific disciplines. Available at https://retractionwatch.com (with posts mirrored on Twitter under the username @RetractionWatch), you can browse alerts that explain what has been retracted recently along with the reasons why. As of October 2018, Retraction Watch also have a searchable database of over 18,000 retractions at http://retractiondatabase.org. You can search by author, title, keyword etc. as with any academic literature database, so your searches can be as broad or as focused as you like. Search results then show not only the usual referencing details that would let you find each retracted article, but also a list of reasons why each article was retracted (e.g. 'plagiarism of article', 'fake peer review', 'duplication of image' and 'error in analysis'). A digital object identifier (DOI) with each result also lets you quickly go straight to the actual retraction notice issued by each journal.

Further questions to ask

This list of points isn't exhaustive. You might find more specific guides for your own particular field that can tell you what to read for in finer detail, or with specific types of article. For example, Trisha Greenhalgh's

How to Read a Paper: The Basics of Evidence-Based Medicine is currently in its fifth edition, and has separate chapters dedicated to individual types of research ('Papers that report questionnaire research', 'Papers that report quality improvement case studies', 'Papers that tell you what to do (guidelines)' and so on). Ask your subject librarian or the staff at your university bookshop for recommendations in your field; both of these jobs require a detailed knowledge of the stock on the shelves and applicability to students studying different subjects.

You'll also find a complementary critical analysis chapter in our previous book, the more general *Writing for Science Students*. There is inevitably some overlap with the concepts we've covered here, but in this book, we've tried to focus more on the aspects that are relevant at PhD level. If you're interested to read more widely, we'd recommend that you try to at least borrow a copy of our earlier book for other step-by-step instructions and examples.

Ambiguity and the realities of scientific research

In the process of critically analysing all of the literature you read in the course of your PhD, you'll probably develop a much deeper appreciation of just how much uncertainty there is around scientific discovery. Working at the forefront of human knowledge, there isn't always an easy answer to a research question. Resources are not unlimited: each project has a budget, an end date, a limited number of brains to think about it, and it will be hit by unexpected setbacks and unexpected results. Different journals will also place different requirements on the manuscripts that researchers may submit to them: word counts, figure counts, level of detail in a methodology, etc. might all be capped for word limit reasons or other stylistic choices that differ from journal to journal – their 'house styles'. We hope that our advice helps you to develop your skills in critical analysis, but we also hope that *doing* the critical analysis brings you a realistic appreciation of just how messy and sometimes problematic the day-to-day work of science can actually be.

CHAPTER 7 **Structuring Your Chapters**

Students often express anxiety over whether the written work they've produced 'looks like' a PhD. Equally, concerns over structure sometimes impede writing. In this chapter, we will think about to what extent the structure of the PhD can be regarded as an extended, chaptered version of the type of lab reports that you have very likely already written in your undergraduate and master's courses.

We'll also consider the kind of variation that can exist within this structure, using the table of contents of a published thesis in order to give you a sense of how structural approaches can differ, and how much control you have over the structural presentation of your work.

We'll also discuss the organisation of content within chapters. The emphasis here will be, as it has been in previous chapters, on writing with the reader in mind: ensuring that that flow of information seems logical to them, and that they always have the context they need in order to understand the work. Chapter introductions and conclusions will be broken down and examined to demonstrate how they can effectively frame the chapter, and how they can act as a means of checking that the information contained therein is relevant and coherent.

Overall, this chapter will look at chapter structure at a thesis-wide level, as well as discussing the internal organisation of chapters. We will explain how while there is an established structure in terms of introduction, methods, results, etc., you should also see structure as a means to effectively convey ideas and argument.

Principles of a good chapter

Just as is the case with smaller units of organisation within the PhD, the sentence and the paragraph, chapters need to adhere to certain principles. They should be:

- unified – the content should all relate to the topic of the chapter
- coherent – there should be a logical progression throughout the chapter
- well-developed – the chapter should be solid and thorough.

Additionally, there should be a logical flow from one chapter to the next that guides your reader effectively through your research, allowing them to plainly see how your work fits into the wider field, exactly what you did, what you found, and how this contributes to the field.

Science students don't tend to tackle their chapters in chronological order.

Insights from a researcher

The methods chapter was the quickest (to be honest, I could have written it long before I finished my lab work) ... I jumped around between chapters a lot because I would hit a block and tell myself to pick something easier to write about. For a long time, I struggled to decide on what should go in the mini introductions/discussions, and what should go in the actual introduction and discussion chapters. I tackled the overall discussion and introduction chapters last.

Insights from a researcher

Don't write linearly. I found I got on better when I jumped around from section to section. For example, when I got stuck with a complicated bit in the literature review, when I lost my train of thought and didn't really know where I was going, I'd highlight it, leave it, and move on to something in another section of the thesis.

Most tend to write the methods section first, finding this to be easiest. Several chapters cannot, of course, be written until the work itself has been carried out, so these can only exist as a skeleton at the outset.

The introduction tends to be the chapter written after the methods section. This is because writing an early draft of the literature review is often the first-year requirement for many PhD students, serving as it does the purpose of familiarising the student with the relevant literature, and encouraging them to think about how their work relates to existing studies.

It's good practice, even though you cannot write many sections until you've carried out your work, to try to work within an overall thesis structure from the outset. The chapter structure doesn't have to be set in stone, but many students find that writing within something that at least resembles a chaptered PhD encourages them to think about flow within and between chapters, as well as breaking the task down, and making it seem less overwhelming.

Typical chapter structure

You are very likely to be familiar with the structure of a typical lab report:

Abstract
Introduction
Methods
Results
Discussion

Within this traditional structure, it might also help you to keep an hourglass shape in mind in relation to a PhD. The introduction starts out giving a broader overview of the current state of play in your field as it pertains to this particular topic. It then narrows to bring the reader to your specific approach, and the work itself. In the discussion section (and conclusion), the focus widens again, moving from the discussion of your data to think about how the work impacts on the field as a whole, and finally to where new directions might take the work.

Within chapters, the introductions and conclusions (or concluding sections of mini discussions) should equally act as frames for the overall chapter, albeit on a smaller scale. The introduction relates the chapter to the overall thesis, often explicitly by cross-referencing to pages or section numbers where you've described other pieces of work you've carried out or observations you've made. It should then tell the reader what this chapter will tell them, and why it's relevant. The conclusion/concluding sections should refer back to this in some way, even briefly. This check will allow you to confirm that you've stayed on track throughout the chapter without losing focus. For your reader, it's a helpful reminder of how they should understand what they've just read and how it relates to the wider piece of work.

In many ways, the PhD adheres to this essential structure. A useful way to start to think about chapter structure might be to identify exactly where and how it differs from a typical lab report.

Variations in chapter structure in the PhD

First of all, the length of the finished PhD thesis is likely to be much greater than that of the average lab report that you produced in your undergraduate or master's studies – this is obvious, as the scope of the project is much wider. On top of this, however, some sections are likely to be significantly more complex.

Introductions

The first place where you might notice this is in the introduction. PhDs deal with a research question that is much more ambitious in scope than any work you're likely to have carried out so far. As such, they require much more context to set up the aims of the study. If you're undertaking interdisciplinary work, and perhaps presenting material that might be wholly unfamiliar for your readers, then this might make the introduction even more lengthy, as additional contextualisation will be required. You might also choose to absorb your literature review within the introduction, creating a chapter which focuses on setting up the background to your work. We see both variations (introduction containing the literature review; separate chapters for introduction and literature review) in our work with students. It's best to be guided by your own sense of what suits your work and your supervisors' advice. It's also worthwhile to look at completed PhD theses from your department, to give you a sense of typical structure here.

The introduction, then, has the task of providing the context your reader needs in order to understand where your work is situated and why you're doing what you're doing. You might find it more helpful to think of it as a 'background' chapter. As it concludes, it provides the jumping-off point for your own work.

Introductions are usually sub-divided into multiple sections, sometimes down to individual pages. The level of detail you choose to adopt in how you sub-divide this is down to your individual judgement on the extent to which specific details merit their own highlighted sub-section within the chapter.

Consider this example:

Table of Contents

1 Introduction 15
   1.1 Crops in a Changing Climate: The Problem 15
      1.1.1 Changes in Productivity and Population 17
      1.1.2 Management Strategies 18
   1.2 Crop Responses to Climate Change-Related Stresses:
      Prospects for Adaptation 19
      1.2.1 Heat Stress Responses 20
      1.2.2 Hydration Stress Responses 23
      1.2.3 Salt Stress Responses 24
   1.3 Aims of This Research 26
   1.4 Project History – MYB64: A MYB Transcription Factor
      Conferring Salt Tolerance 27
      1.4.1 Mutant Search Strategy: Activation Tagging 27
      1.4.2 MYB64 Functional Characterisation 29
      1.4.3 MYB64 Transgenic Overexpression Line 31
   1.5 Project History – The Small Heat Shock Proteins
      (smHSPs) 31
      1.5.1 smHSPs in Plants 32
         1.5.1.1 smHSP Sequences, Structures and
               Functions 33
         1.5.1.2 Regulation of smHSP Activation 39
      1.5.2 MYB Genes in Plants 40

This is fairly typical in terms of level of detail. The writer could have taken previous sections to even deeper levels, but they decided that this was not warranted, and have made the decision based on what they felt served the work and guided the reader.

Structure within the introduction is ultimately down to the writer. You might feel that a chronological approach is the most effective way to explain the background to your work, charting developments in the field. If you are carrying out interdisciplinary work, then you may choose to let this influence your structure and make sub-divisions by discipline. Alternatively, you may prefer that a thematic approach is the best way to tackle the structure, sub-dividing the material by topic, and drawing the reader's attention to those aspects of the topic which are most relevant to your project.

If you look back to the introduction outline on p. 89, you will see that this is the approach that this author took.

Their study analyses a specific type of gene, known as a transcription factor, and the role it plays in heat and salt tolerance in crop plants. As such, they've chosen to structure their introduction by first presenting the problem. They then move on to specifically addressing how crops might respond to specific stresses. Within this second section, we have sub-sections for each type of stress that is of interest: heat, hydration and salt. There's a single section to detail the aims of the study and then an overview of the history of the study, again highlighting those aspects that the writer sees as most relevant in relation to the current study.

You can see here that the writer could have made this introduction more or less detailed, if they had preferred. For example, the section on the response to heat stress could have had more sub-sections, but the writer has opted not to take this approach. Equally, you have the same level of freedom to make structural decisions within chapters based on your understanding of the work and how it is best presented to enable the reader to follow it without confusion.

A similar approach can be taken in other chapters, such as the method chapter. You should be guided by practicality. Structure the chapter in the way that best enables the reader to follow what you have written. If you have difficulties remaining distant enough from the work to remain objective, then ask someone else if they would be willing to read it for you.

We would emphasise that the type of structural approach above is common, but by no means the only structural approach. Again, we would advise you to be guided by the shape of your own research, and the norms within your own subject area. The excerpt is from a life sciences PhD thesis. The approach taken in a thesis in physics or computing science might look quite different. Equally, there can be a great deal of variation *within* each of these subject areas. Look at existing theses to give you a sense of both what's expected and the options open to you.

Body chapters

Your structure, to a great extent, is likely to be driven by your work. For example, if your study uses a novel method, but applies it in five very different contexts, then you might find that it makes sense to have five chapters (one per context) which have their own mini-literature reviews at the outset of the chapter, and five specific mini-discussion sections within each chapter. This would enable you to ensure that you discuss the use of the method in these contexts in detail, making sure the reader was fully informed, but would also allow you to leave your general discussion section for a broader discussion about the utility of this method, comparing your five different contexts and bringing in details as necessary.

This is not to say that you couldn't, if you preferred, keep all of your literature review content at the beginning of the thesis, and have one very long discussion chapter at the end. Bringing together the data and discussion points from five very disparate contexts is likely to be difficult to write, though, and might not serve the data well. It also places quite a heavy burden on the reader, who will have to bear in mind the information they have taken from each chapter while they read your discussion. However, you might feel a comparison of every detail in one coherent discussion section is the best way to bring the work together. It's really a matter of what is easiest for you to write, what is easiest for your reader to understand and what serves the work best. There are no hard rules about what must or must not be done in this aspect.

It should be clear to you how the chapter is relevant to the overall study. Each chapter should have some kind of introduction which also makes this clear to the reader, and foreshadows what will be covered in the chapter. Depending on the length and content of the chapter, this introduction could vary from one to multiple paragraphs. The length of the introduction itself is less important than whether it does its job.

Concluding sections should make clear to the reader the purpose of what they've been asked to read and how they should understand it in the context of study.

Writing a short description in your notes of what each chapter does can help you to retain focus within that chapter, and ensure that you keep the reader's perspective in mind. Does the chapter explain something? Contextualise? Summarise? Discuss? Illustrate? Individual chapters are often too complex to be broken down into this kind of one-verb-summary (except, perhaps, the methods section), so think of all the different purposes the chapter is attempting to serve.

It might begin with contextualising what it's doing. It might then describe a method in detail. It could go on to discuss the outcomes of the work that was carried out, compare and contrast expectations versus findings, and then speculate about future applications. Thinking of the purpose of various sections of the chapter can help you shape it and ensure that each section effectively does what you intend it to do. You could colour code the document as you work on it to remind you of what you should be aiming for in each section.

Again, we would point out that structural approaches within chapters, and the typical number of chapters within the thesis itself, can and will vary depending on your discipline. As such, analysing the structures of published PhD theses in your subject area is an excellent way for you to begin thinking about your own structure, or perhaps to open up possibilities if you're having difficulties thinking about how to tackle the task. Most universities have online collections of their past students' PhD theses. Glasgow's, for example, is http://theses.gla.ac.uk. A possible approach when analysing existing theses is as follows.

> First, read the research question. Ask yourself what topics it will likely span and, subsequently, what chapters and sections might seem immediately sensible to you, based on the question.
>
> Now look at the chapter page. Does its structure reflect your expectations? If you read over it, does the structural approach aid your understanding? If not, what type of structure would have helped? More or less detail? Specialised discussions in each chapter to help you absorb information before you moved on?

Look at other theses in the same subject area. Is this thesis atypical? Fairly conventional? Can you begin to get a sense of the structural norms and standards within your subject?

Try to extrapolate your own set of guidelines for what makes a well-structured piece of work, and then consider your own thesis, and ask whether you are working to those guidelines.

Lastly – remember that your structure is not set in stone throughout the writing process. The structure reflects your research and how you choose to communicate that research to your reader. Your ideas on that are likely to shift throughout the process, and you should remain open to changes: a responsive structure is the sign of openness to ideas and a reflective researcher.

CHAPTER **8** **Writing About the Literature**

Our previous few chapters have covered the collection, organisation and evaluation of other people's work. If you've followed this advice, you should now be in the fortunate position of benefitting from a well-curated library of sources, each of which will have annotations in some form regarding your thoughts on the strengths and weaknesses of each paper, and the inter-relationships between them. Now, it's time to deploy those sources in a way that supports your central argument, and in a way that makes you sound like a professional scientist.

We've noticed that many of our students don't think of themselves as writing an *argumentative* piece of writing. An argument, though, is simply a statement of what you believe to be true, supported by your reasons, supported in turn by evidence. Scientific evidence is collected in as objective a way as possible, and we assume every reader should therefore be able to interpret it in the same, correct way. As we saw with the previous chapter, though, this isn't always the case.

How many sources do I need?

This is a question we hear a lot from our students. Sadly, there isn't an easy answer. We've heard some staff say 'I expect one reference for every 100 words', and we've read professional literature that has only a couple of dozen references for a whole article. This is because the answer varies by discipline. Some fields of research are very active, in which case you would easily find an abundance of new articles published each year. Other fields are more specific, niche subject areas, in which case the research community that produces its articles would be much smaller, and much less prolific.

The nature of some disciplines is that lots of individual papers might be cited in order to build up a clear overall picture, while other fields might be based around deep, qualitative analysis of a few rich sources of information. Since the type of work you'll be doing in your PhD is probably reflective of the type of work the existing academics are doing, you can likely expect that your ratio of citations to total word count would be about the same as what you find when you read theirs.

Of course, as always when dealing with technicalities that could appear in rules and regulations, check with your supervisor or your university for their own opinion on this particular issue.

How often should I use sources?

This is a subtly different version of the previous question that we sometimes hear. Rather than dealing just with a total reference count, this question is slightly deeper, as it deals with how the piece of writing should be constructed. Again, there's no easy answer. 'As frequently as is necessary to support your argument.'

Which sections of my thesis should include references?

A sensible question to think of, and a topic that's interesting to think about if you haven't. It's plausible that you'll have references in every section, but some sections will have many more references than others.

Your **introduction** is where you contextualise your research. This is where you'll be relating your work to that of others, and explaining where the gap in our knowledge is, so that your reader will understand why you're designing your research in the way that you are.

Your **methods** section will have references too, but definitely not as many as your introduction. If you've adopted (or adapted) someone else's method, you should acknowledge them as the creator. This might be by writing 'X was carried out according to the protocol by Smith and Jones (2003): …', where you go on to explain what that method was. If you've modified it, you might write 'X was carried out according to a modified version of the protocol by Boyle and Ramsay (2018): …' In both cases, you should then write out the method in full.

This differs from what you'll read in some journal articles, where the authors will stop at the acknowledgement and leave their readers to go and find that original source in their own time. This is done in some journals because of a pressure to keep within a certain word limit, which in turn is done to encourage a higher readership and increase the reach, and therefore the impact, of a journal's articles. The full details might be published as appendices, downloadable from the journal's website.

It isn't something we'd recommend you adopt in your PhD, though, because your primary goal is to show the examiners everything you did. A classic question at viva is 'tell us about what's in chemical buffer A?', or 'What does "according to the manufacturer's instructions" mean? Tell us about the steps, and what purpose they each served'. The more you can write about this in your thesis, the fewer difficult questions you'll be likely to receive later.

Your **results** section will primarily include your own findings, but it might still include references to the work of others if you're either drawing comparisons between your work and theirs, or if you're describing a variation on a method.

It's a procedural error for you to start to *discuss* your findings in the results section of your work, so in 99% of cases, you shouldn't need to cite someone else's work. The exception to this might be if you make a discovery that doesn't fit with your expected project plan, and which causes you to change what you do next. Your reader will be following the timeline laid out in your results section, and they'll need to understand the rationale for such a significant change in approach, so you would describe those results as surprising and describe the impact this had on what you planned to do next. Even in this case, it's unlikely that it'd be appropriate to start comparing your findings with those of others in order to justify the change, but, in the spirit of any good scientist, we aren't foolish enough to say 'never', 'always' or 'impossible' about the prospects of *anything*, so we'll say there's a very slight chance that the change is so outlandish that you and your supervisor feel a citation to someone else's work would provide that necessary context, even in your results chapter.

Your **discussion** section should contain many references, second only to your introduction. Your job in this section is partly to remind the reader of the context you introduced at the start, so some of those citations will be there for this reason, and partly to put your novel findings into context with wider scientific knowledge. Some of the articles you cite in your conclusion will be unrelated to the topics you mentioned in the introduction simply because the nature of science is that you can't predict where your research will take you. You should be talking about how your findings match, contradict or otherwise complement this body of knowledge. Doing so means you weave your research deeply into that academic conversation, rather than simply sitting on top of it without much integration.

What are the different reasons for citing others' work?

There are two main ways to classify the reasons for referencing the work of others: academic integrity, which includes avoidance of plagiarism; and joining an academic conversation that goes beyond the limits of your own lab group.

To give credit

Fundamentally, it's important not to attempt to gain academic awards on the basis of using someone else's work. Let's examine what this means.

Citations are the markers we use in writing to show who created an idea, a conclusion, a figure or a piece of communication. In the world of television and film, the only way that a relatively anonymous contributor (like the lighting director) can prove they worked on a production is to have their name included in the end credits. This way, when they're applying for work on their next project, they have some evidence that they contributed to that work. Similarly, in academia, citation in the works of others is the way that a line of accreditation can be maintained back to the original source.

This allows readers at some point in the future to trace a chain of citations from a new article all the way back to the first time that the idea or the finding had been published. More than this, it also allows the reader to take the name of the author and to search for the rest of their publications. You'll be doing a literature review for your thesis, so you'll recognise that this is an extremely important and extremely useful attribute, as it means you don't need to rely on a key word search to identify *all* of the articles that are relevant to you and your research.

> ... *for ideas*
> Ideas are intellectual property. If another researcher formulated a hypothesis or designed an experiment in a particular way, that counts

as their own work. If you want to use those ideas as foundation for your own research, then you need to make sure that you credit the originator of those ideas in the same way that you would cite them for their words.

... for analyses

Work carried out by other researchers will take place in a variety of physical settings. It's easy to recognise that lab work counts as research, and that any work you choose to show in your thesis from someone else's physical research needs to be credited to that person. If you find a statistical analysis, though, does this count as work in the same way? Anyone could take the original data published by the authors and carry out their own numerical analysis of it.

The fact that the work *has* already been done, however, means you mustn't present, use, adapt or otherwise incorporate the results of that analysis without crediting the person or people who first published it.

... for conclusions

This situation is essentially the same as the ideas example mentioned above. If you write about an article in your thesis and you want to tell your examiners what the results of that piece of research *mean*, and if you concur with the conclusion of the original authors, you aren't coming to an original conclusion. (Unfortunately, you don't get any credit for contributing to a scientific conversation by simply repeating what everyone else has already said!)

If you want to report a conclusion made by the authors that you agree with, you would fold it into your paragraphs like this:

> ... which resulted in a five-fold increase. As concluded by the authors, this shows that ...

... for words and figures

This is the more familiar situation when you need to cite someone's work. We do this as scientists for reasons of academic integrity, but, interestingly, copyright law recognises text as intellectual property in a very similar way. When you put words together in sequence (and that sequence is new), copyright law regards this as an original work authored by you. Copyright is a protection that doesn't need to be applied for; it is automatically granted instantly at the point when a work is created. (This differs from, for example, trademarks, but the copyright and trademark symbols are often seen in the same context, so they are often confused.) It doesn't only apply to formal publications. Copyright applies to websites and even your own personal diary just as much as it applies to articles and books.

The fact that the law recognises sequences of words as intellectual property only serves to underpin our argument that you must credit an author when you use the same piece of language they created in their article. Not to do so would constitute academic dishonesty.

Legalities aside, there is another (perhaps bigger) reason, and it's specific to the notion of being a student.

As someone who is, by definition, not yet qualified at the level where you're currently working, it's important that you show not just the finished product of your work, but that you can prove your skill to an examiner. This means that while you might completely understand everything you read in the literature, you have to demonstrate what that means to you and how those pieces fit together. This means putting the original text into your *own words*.

We've been asked many, many times why this is so important to academic institutions. There are several reasons.

Reason 1: Using your own words shows what you've understood the original source to *mean*.
- You'll read each article with a question in mind. What experimental procedure are you looking for guidance on? What knowledge gap of yours are you trying to fill? It's highly unlikely that an article concludes in such a way that neatly answers this in words that you could also transcribe directly into your thesis. Your job is to extract a nugget of meaning that's relevant to you, and this will usually involve articulating the crucial piece of meaning in words that are more specific and more relevant to your situation, and often less numerous.

Reason 2: Using your own words allows you to weave the literature into your own argument.
- Your literature review (uses of literature) shouldn't be a report of everything you've read. It should be a *discussion* of what knowledge exists, and how that all relates to the project you're carrying out. You can't hope to do that by simply pasting together pieces of

text from other people, and inserting your own connective phrases between the words of others will only go so far to making this work.

Reason 3: It makes your degree worth something.
- If your university accepted the argument we described above – that it's possible to understand something without changing the words of it – then it leads to a difficult task of assessing when this is the case and when it isn't. How you would differentiate between someone who intentionally plagiarised many short pieces of text to create something that looked like a PhD thesis with a minimum of effort, and someone who did a thorough job of understanding the literature yet used all of the original wording? Would you be able to differentiate at all? If your university was known for accepting this practice, it would devalue your degree.

The same concept applies to the use of figures, diagrams, tables and other visual elements you've incorporated from sources other than your own work. This applies *even if you modify them*. You might choose only to include one section of a figure, perhaps because it's a multi-panel figure showing the results of four or five experiments at once. If you only want to show one of those panels, you must acknowledge the source and, crucially, you must also say that the figure has been modified. For more details on this, see the section titled 'How to use the works of others without breaking any rules'.

To help your reader locate the original source

You'll no doubt have found literature through a combination of searching academic databases and following references within articles because they sounded interesting and relevant. You'll recognise the importance, then, of making sure your references are complete, including all of the information pertinent to that source (the usual set being author name(s), date of publication, article/chapter name, journal/book/online resource name, volume and issue/edition number, page number(s), place of publication (for books) and DOI (digital object identifier, for journal articles – helpful but optional, as not every article will have one)). For a fuller discussion of these elements, see Chapter 4, 'Finding Literature'.

Make sure to expand the titles of any journals you find mentioned in abbreviated form (e.g. 'Acc. Chem. Res.' should be 'Accounts of Chemical

Research', and 'Biosens. Bioelectron.' should be 'Biosensors and Bioelectronics'). This avoids causing the reader the hassle of guessing or searching for the abbreviation before being able to locate the journal. (Remember, 'your reader' in this case also means 'your examiner', and if your thesis has other problems then you don't want to further compound them further by presenting references that are more difficult to follow than they need to be.)

Plagiarism

The consequences of plagiarism accusations at university are severe. You are likely to have had this impressed on you throughout your undergraduate career, but the consequences are heightened when working at PhD level, where plagiarism is classed as serious research misconduct. We don't expect that anyone reading this is intending to commit plagiarism, but you need to know what it looks like in reality so you can make sure you don't do it accidentally. The worst-case scenario is probably that you graduate with your doctorate, you find a job where you put your PhD to work, you perhaps move to a new part of the world, you establish a reputation as a professional, and then someone discovers a piece of work in your thesis that matches the work of someone else. The university would no doubt hold an investigation into the accusation and, if it was upheld, your doctorate could be withdrawn. Your professional position as an academic would then, obviously, be untenable.

Slightly less extreme, and far more common amongst instances of plagiarism, is the chance that someone discovers a passage of high similarity while you are still undertaking your degree. This is most likely to happen at your viva when your external examiners go over the document in great detail, and perhaps they decide to read some of the articles you've cited. Being experts in your field, there's also a chance they've already read those articles themselves. At this stage, by handing your thesis in, you'll have already formally submitted a piece of work to the university for assessment. This carries the assumption that you claim the work to be entirely your own unless otherwise stated at the appropriate point within the thesis itself. In fact, this assumption is made explicit if and when you sign an accompanying declaration of originality that your university might use. This means that any inappropriate similarity between your work and that of others will be treated as an attempt to gain credit by plagiarising, almost certainly triggering disciplinary procedures.

We understand that not every instance of high similarity in a piece of submitted work is the result of a deliberate attempt to commit academic fraud; sometimes, there's just very little you can do to rephrase the work of the original authors. So, if this is the case, why do we insist that you use your own words at all?

How to use the works of others without breaking any rules

Broadly speaking, there are three ways you could incorporate someone else's work into your own: quoting, paraphrasing and summarising. We've gone into detail on this in our previous book, *Writing for Science Students*, which covers writing at a level for those who have never written for a scientific audience before, so we won't spend long on the difference between the three approaches here. Instead, we'll summarise the main points:

> **Quoting:** Using someone else's words directly, within quotation marks, and citing the source. Not often used in science; only really appropriate if it's absolutely crucial to retain the original wording, for example if you are quoting from a scientific policy or legal document, healthcare system guideline, etc.
>
> **Paraphrasing:** Using someone else's sentence, or even their paragraph structure, but changing the words for synonyms. Extensive paraphrasing becomes very close to plagiarism, and, at the very least, is quite lazy practice.
>
> **Summarising:** Incorporating the part of someone's work that you think to be most relevant to the point you are trying to make; usually involves using far fewer words than were in the original.

Rather than repeating ourselves unnecessarily here, you can find a more extensive exploration of these distinctions if you wish in our previous book, *Writing for Science Students*.

Remember that copying is not simply about the use of the words; writing is also about creating structure and a pattern of argumentation. It's possible to copy the essence of someone's work without using the actual original wording, and it's certainly also possible to copy the underlying intellectual ideas about the science being described.

If you're including a figure, table or any other type of element that doesn't form part of the main body of your thesis text, you must also cite the source for that. You would usually do this at the end of a figure caption.

If you wanted to be very thorough, you could go an extra step to make it clear whether the figure alone came from the external source, or whether the figure *and* the caption are someone else's work.

If you're modifying a figure at all, you must tell your reader you've done so. This is mainly so you don't misrepresent the work of the original author. Imagine a scenario where you find a very complex diagram of an experimental set-up complete with many labels, and you want to show this to your reader, but you only need them to see a small fraction of those annotations. You might be able to mask out the unnecessary labels in a photo editor, or you might choose to redraw the diagram from scratch incorporating only the parts you think are relevant to you. If you cite the original author next to either of those, you would misrepresent what they originally produced. In such cases, make sure to include the phrase 'adapted from…' or 'modified from…' before you include the usual citation.

Citing articles within articles

You might sometimes find that you want to include a finding that didn't actually come directly from the article you have in front of you, but from another article that this author has read. You are then *two* layers away from the original source, rather than just the usual one. The easy, tempting option is to paraphrase the intermediate author who reported the findings of the original one, cite the original author's name and move on. However, you then risk misrepresenting the actual truth of the science. What if the intermediate author summarised only part of the original author's research, and what if that extra context would mean the finding actually isn't as relevant to you as you first thought? What if the result nicely complemented the work you're doing, but the methodology used by the original author was totally unlike the one you are using, and therefore not comparable? What if the intermediate author actually *misunderstood* the original findings, and was totally wrong in their conclusion? Or what if, by virtue of the fact that each stage of the process involves changing the wording further from the original text, you believe you've done everything correctly, but the synonyms and variations on phrasing you and the intermediate author have used actually mean something different from the original text?

In short: never cite an author's work until you've laid eyes on it yourself and you've been able to critically assess its value.

That said, in the extremely rare case that the original article is so old that it is out of print and has never been digitised, or if it's in a book that your library simply doesn't have access to, your last resort is to cite *both* the original and the intermediate authors. This tells your reader that you haven't been able to read the original, and implicitly lets them know that there may be errors in your interpretation, but that your work has been done in good faith.

Self-plagiarism

It's important that you are also aware of **self-plagiarism**, a subject which is not often covered during undergraduate studies.

Self-plagiarism occurs when an author reuses previous work. This, as you can imagine, can also have serious legal repercussions, since replicating work from a previously published study (even your own) has legal repercussions in terms of breach of copyright. 'Work' doesn't only mean an entire article. Reusing sections of text, data or images would still be unacceptable.

When deliberate, this type of misconduct tends to occur in the context of publication, where researchers who are keen to build up their profile will submit superficially altered versions of the same work to different journals.

This type of plagiarism can occur within the thesis, however, and is often the result of genuine confusion over ownership of work. Again, this can take a variety of forms. Reuse of images, data and wording would all count as self-plagiarism. If you are quoting work that you submitted for another qualification, then you **must** fully reference that work as you would another author.

If you plan on publishing any sections of your thesis during the PhD, then you should also make yourself fully aware of the relevant copyright regulations. Practice differs widely in regards to this. If your institution has the option to produce a PhD thesis by publication, then there are likely already clear guidelines in place. Similar guidelines are likely in existence if your institution encourages students to publish sections/chapters of the thesis during the PhD (as opposed to a formal PhD by publication route). You should, as early as possible in this process, familiarise yourself with guidelines within your institution and talk with your supervisor. It is your responsibility as a professional to ensure that you are fully informed and conduct yourself appropriately.

Referencing styles

There are two main style types in science: numbered and alphabetical. The third main category is footnotes/endnotes, but this is typically only used in the arts and humanities. The most popular numbered style is known as Vancouver, and is particularly popular in medicine. Its benefits are brevity and a cleanness of presentation, as numbers leave the sentences virtually uninterrupted to the eye. The most popular alphabetical style is known as Harvard, and is more popular broadly across the sciences. Its benefits are that you can immediately tell both how recent a source is, and you can easily recognise where the same author is mentioned several times without having to check the reference list at the end. This also allows you to quickly notice works by the same lead author, which allows you to build your sense of the authors who publish extensively in your field.

As with the three main ways to incorporate someone else's words, which we mentioned above, we have covered this topic in greater detail in our previous book, *Writing for Science Students*.

CHAPTER 9 Structuring Your Sentences

This chapter will start by explaining **sentence structure** at a depth that you've perhaps never considered before, and will also discuss how a solid knowledge of sentence types and the roles they can perform gives you a fine level of control over the presentation of your work. This, in turn, helps to control your readers' interpretation, allowing you to communicate information as effectively as possible, as well as avoiding the potential for ambiguity and misinterpretation.

The second half of the chapter will give an overview of **punctuation**. Adopting the same approach as the section on sentence structure, we'll avoid simply reciting the rules of punctuation, and instead we'll let you see how the rules work in practice, how you can use them to your advantage, and how you can avoid the most common errors we see in student writing.

Sentence structure types

Mastering effective sentence structure is not something that is near the top of most science students' list of priorities when starting to write their PhD thesis. However, an understanding of sentence structure is enormously useful in clearly and concisely conveying exactly the kind of complex, dense information that you will be dealing with in your work. Additionally, uncertainty over how sentences should be structured can often result in work that's needlessly difficult to read, with novel, interesting points buried and lost in convoluted structures. At its worst, your understanding and contribution can both be obscured, leading to misinterpretations of both your meaning and your abilities.

Here, we will present sentence types (simple, compound, complex, compound/complex), and discuss how they can be put to work for the writer.

Simple sentences

Simple sentences contain only **one** clause. This clause can stand alone, and the reader requires no additional clauses for the simple sentence to make sense. It can be known as a **main** or **independent** clause.

- We carefully noted the results.
- You read books.
- The collective decision was eventually made to abandon this initially promising approach.
- Transgenic studies have repeatedly shown that overexpression of various plant and fungal smHSPs and HSPs has improved the thermotolerance of the host plants at critical temperatures (Ramsay, 2014, p. 21).

Notice that the length and complexity of the sentence can vary greatly. Simple sentences need not always be very short. The common feature here is that they consist of only one clause, expressing one idea without any additional or qualifying information.

As such, simple sentences are particularly effective when you want to draw attention to one particular idea or fact and impress it on the readers' minds. As such, if you have made an observation that you feel is particularly astute, or you are presenting an original finding, or an important conclusion, then you might want to consider using a simple sentence.

Compound sentences

Compound sentences contain **two independent clauses**.

- Some researchers work well alone; other researchers work well in a team.
- The first set of results seemed especially promising, but the second set presented significant problems.
- Such responses to drought may include overt phenotypes like growth arrest; this is advantageous as the cessation of growth means that any metabolic activity can be focused on avoiding or ameliorating the effects of the stress (Ramsay, 2014, p. 23).

Notice that as they are independent clauses you could, if you preferred, make them separate sentences.

- Some researchers work well alone. Others work well in a team.
- The first set of results seemed promising. The second set presented significant problems.
- Such responses to drought may include overt phenotypes like growth arrest. This is advantageous as the cessation of growth means that any metabolic activity can be focused on avoiding or ameliorating the effects of the stress.

If you want to emphasise that the sentences are linked by meaning, and that they flow on easily from each other, then it's likely that you would prefer to write them as compound sentences, emphasising this connectedness for your reader. If, however, you feel that the information is particularly meaningful, and you want it to stand out, then you might choose to present the information as two separate simple sentences.

As we said, an understanding of sentence structure is primarily about giving you complete control over how you communicate information and express your ideas; there is no right or wrong answer here.

Compound sentences are also particularly useful if you want to draw your readers' attention to a contrast. Consider again the example above.

Some researchers work well alone; other researchers work well in a team.

Using a semi-colon alone draws attention to the contrast and provides a sense of balance to the sentence. You could choose to use a comma and a conjunction instead.

Some researchers work well alone, but other researchers work well in a team.

This is perfectly grammatically correct. In terms of style, though, it doesn't have quite the same force as the first sentence. It could also be argued that the introduction of 'but' before the second half of the sentence has an impact on how the reader interprets the information. There is now perhaps more focus on the second half of the sentence. Is the fact that these researchers work better in a team notable? Is the fact that their working style differs from the researchers in the first half of the sentence contradictory to some widely accepted way of thinking or somehow difficult? It's a small change but, as you have seen, it does have an impact on how the information is conveyed and understood.

You will notice, if you look at these examples, that the compound sentences are linked differently in terms of punctuation. Two of the sentences use a semi-colon to connect the two independent clauses. The sentence in the middle uses a comma and a conjunction. We'll talk about these in more detail later in the chapter. For the moment, simply note that these are the appropriate ways to link two independent clauses when you want to form a compound sentence.

Complex sentences

The next type of sentence you might use in your writing is a complex sentence. This consists of an independent clause and a dependent clause. As with simple sentences, the length and complexity of these sentences can vary, but the essential structure remains the same.

- As the instruments were unavailable, an alternative approach was selected.
- If we are to move towards an integrated model of the stress responses elicited by a changing climate, it is important to first of all recognise what has been discovered about the individual components thus far (Ramsay, 2014, p. 19).
- There are a number of different contexts in which this approach might be beneficial, such as the aviation industry.

You can see here that the dependent clauses, the clauses that cannot stand on their own, exist to provide the reader with additional information. While the main clause delivers the information that is the key point of the sentence, the dependent clause usually provides either context or a caveat.

As such, if we try to separate these sentences in the same way as we did with the compound sentences, we can easily see that the dependent clause cannot stand as a sentence on its own.

- As the instruments were unavailable.
- If we are to move towards an integrated model of the stress responses elicited by a changing climate.
- Such as the aviation industry.

This is also the case with compound/complex sentences, a closely related sentence type.

Compound/complex sentences

Compound/complex sentences contain two independent clauses, and at least one dependent clause.

> The cost of the materials presented difficulties, but the experiments were continued because time was an important factor.

Let's analyse the elements of the sentence in more detail.

- The cost of the materials presented difficulties – independent clause.
- The experiments were continued – independent clause.
- Because time was an important factor – dependent clause.

Notice that the dependent clause here is not indicated with any kind of punctuation. As a dependent clause, though, it can still be removed and leave a coherent sentence.

> The cost of the materials presented difficulties, but the experiments were continued.

You could equally have a compound/complex sentence with the dependent clause indicated through punctuation.

> Smith, who advocates this approach, rejects any alternatives; however, Gerstmann embraces them.

Again, let's break down the individual elements within the sentence.

- Smith rejects any alternatives – independent clause.
- Gerstmann embraces them – independent clause.
- Who advocates this approach – dependent clause.

You can easily understand, by looking at these examples, why complex and compound/complex sentences can be especially common in academic writing. We are usually delivering complex information and ideas, requiring a great deal of context and many caveats. As such, dependent clauses are often necessary in our sentences. They are perhaps especially common in the PhD, where writers are writing defensively, invested in convincing the reader that they have fully understood every aspect of the topic, occasionally overproviding supporting and additional information.

How to use sentence structure to your advantage

As you can see from the examples above, all of these sentence types have their uses. The trick is knowing how and when to use specific types for maximum effect. For example:

> As we had initially hypothesised, our key finding, after running the final test, was that levels of magnesium in the bloodstream dropped after the drug was administered; subsequent studies should investigate this further.

We have a lengthy compound/complex sentence here, with multiple embedded clauses. Nothing about it is incorrect. Their key piece of information, though, that magnesium levels dropped, is buried in the middle of the sentence, and doesn't exactly stand out. Given that this was apparently their key finding, this isn't ideal. Consider this alternative:

> Our initial hypothesis as to the effect of the drug was proven correct. Levels of magnesium in the bloodstream dropped. This is the key finding of this study, and should be investigated further in subsequent studies.

You can see here that by creating a separate simple sentence, this information now stands out for the reader.

You could, if you preferred, retain the compound/complex structure – as we said, it's not incorrect. However, if you do prefer to retain this structure, then consider reordering the sentence to make sure that the important information isn't lost in a subordinate clause between two lengthy clauses. You could also aim to be more concise:

- As we had initially hypothesised, our key finding, after running the final test, was that levels of magnesium in the bloodstream dropped after the drug was administered; subsequent studies should investigate this further.
- The key finding here proves our initial hypothesis: magnesium levels in the bloodstream dropped after administration of the drug. This should be investigated further.

Considering sentence structure at this level of detail is unlikely to be something you would do in the first or second drafts, and is probably best avoided when you're trying to get your first thoughts down on the page. It's more likely to be something you would want to think about in the editing stages, when your mind is less focused on your own understanding of the content, and more focused on your product draft, where you can best facilitate your readers' understanding of your work.

Let's use another example to give you a better sense of how to do this in your own work. Again, there are no errors here. This is just to give you a sense of how malleable the text is once you have a strong grasp of sentence structure.

> Salt is an abiotic stress factor which presents a longer-term rather than an immediate survival problem, and the experiments on halotolerance reported in this thesis were designed to determine responses over a stress period of days and weeks. High levels of salt in the growth medium will lead to high levels of accumulated salt in the plant tissues. Various mechanisms operate to counteract this influx, such as sequestration (Zhu, 2001), efflux via antiporters such as SOS1 (Yang et al., 2009), apoptosis (Price, 2005), and exclusion at the exterior surface of the root, but the inevitable consequence of a sufficiently high concentration is that

some salt will end up in the cytosol and subcellular compartments (Han et al., 2015). The effect of this is to increase the ionic strength of the cytosol, leading to altered secondary and tertiary protein structure. Proteins affected by this will be subject to the same threat as proteins in a heat-stressed cell: denaturation. The role of smHSPs and other chaperones in a salt-stress response thus becomes clearer (Ramsay, 2014, p. 207).

Salt is an abiotic stress factor which presents a longer-term rather than an immediate survival problem. The experiments on halotolerance reported in this thesis were designed to determine responses over a stress period of days and weeks. High levels of salt in the growth medium will lead to high levels of accumulated salt in the plant tissues; however, various mechanisms operate to counteract this influx, such as sequestration (Zhu, 2001), efflux via antiporters such as SOS1 (Yang et al., 2009), apoptosis (Price, 2005), and exclusion at the exterior surface of the root. The inevitable consequence of a sufficiently high concentration, however, is that some salt will end up in the cytosol and subcellular compartments (Han et al., 2015), increasing the ionic strength of the cytosol, leading to altered secondary and tertiary protein structure. Proteins affected by this will be subject to the same threat as proteins in a heat-stressed cell: denaturation. The role of smHSPs and other chaperones in a salt-stress response thus becomes clearer (adapted from Ramsay, 2014, p. 207).

You can see where changes have been made. To increase the impact of the topic sentence, the opening sentence has been changed from a compound sentence to two simple sentences.

- Old – Salt is an abiotic stress factor which presents a longer-term rather than an immediate survival problem, and the experiments on halotolerance reported in this thesis were designed to determine responses over a stress period of days and weeks.
- New – Salt is an abiotic stress factor which presents a longer-term rather than an immediate survival problem. The experiments on halotolerance reported in this thesis were designed to determine responses over a stress period of days and weeks.

The third sentence and fourth sentence have now been combined to form a compound sentence which reflects the close relationship between their content.

- Old – High levels of salt in the growth medium will lead to high levels of accumulated salt in the plant tissues. Various mechanisms operate to counteract this influx, such as sequestration (Zhu, 2001), efflux via antiporters such as SOS1 (Yang et al., 2009), apoptosis (Price, 2005), and exclusion at the exterior surface of the root, but the inevitable consequence of a sufficiently high concentration is that some salt will end up in the cytosol and subcellular compartments (Han et al., 2015).
- New – High levels of salt in the growth medium will lead to high levels of accumulated salt in the plant tissues; however, various mechanisms operate to counteract this influx, such as sequestration (Zhu, 2001), efflux via antiporters such as SOS1 (Yang et al., 2009), apoptosis (Price, 2005), and exclusion at the exterior surface of the root.

On top of this, the clause about 'the inevitable consequence' has been removed from the end of the fourth sentence and placed at the beginning of a new sentence which incorporates the sentence that follows.

- Old – but the inevitable consequence of a sufficiently high concentration is that some salt will end up in the cytosol and subcellular compartments (Han et al., 2015). The effect of this is to increase the ionic strength of the cytosol, leading to altered secondary and tertiary protein structure.
- New – The inevitable consequence of a sufficiently high concentration, however, is that some salt will end up in the cytosol and subcellular compartments (Han et al., 2015), increasing the ionic strength of the cytosol, leading to altered secondary and tertiary protein structure.

Again, we'd point out that there aren't any errors in the original paragraph. What we want to demonstrate is that there is room for someone who understands how to structure sentences to make changes that can subtly change the nuance and emphasis within the paragraph, which in turn has an impact on the readers' comprehension.

It can sometimes be difficult to apply this level of objective analysis of structure to your own work, as you are very close to the content. You might find it helpful to continue to practise this skill on text from published theses and journal articles before you apply it to your own writing.

Punctuation rules and common errors

Commas

Commas are a versatile piece of punctuation, with six main uses in your writing – which we will detail here. However, this versatility means that they tend to either be scattered about indiscriminately in an effort to punctuate overly long sentences, or that they're omitted when they need to be present. For example:

> Although the finding was obtained the work was delayed due to the difficulty in locating suitable specimens.

If we think back to the examples above, we can see that we have a complex sentence.

- Although the funding was obtained – dependent clause.
- The work was delayed due to the difficulty in locating suitable specimens – independent clause.

In complex sentences, it's important that you clearly indicate to your reader the division between the independent and dependent clauses. If you fail to do this, you can create confusion as to the main point of the sentence, which can have an impact on your readers' understanding of your work. This is more apparent if we read a more complex example.

> Transcript profiling produced a list of 'upregulated' sequences of which a significant proportion were previously shown to play key roles in abiotic (and, to an extent, biotic) stress responses, consequently the robustness of these responses in the transgenic lines was investigated by qPCR under heat stress, and the phenotype of the plants was characterised in response to various stress regimes. The findings implicate MYB64 in the regulation of a wide range of stress responses and as plants are unlikely to encounter stress factors individually outside of the controlled conditions of a laboratory these findings highlight the importance of considering such stresses in concert rather than isolation (adapted from Ramsay, 2014).

What is the main point of this sentence? Is it the roles played by upregulated sequences? Or is it the robustness of the responses? Not only has the writer overloaded the reader with too much information in each sentence, but they have also 'comma spliced' – tacked main clauses together with commas. This makes it very difficult to figure out which information is of primary importance, and which is simply useful supporting information.

Your reader might be able to deal with one or two sentences like this if they have the time and patience to tease out your intended meaning, but you can clearly see how too many sentences like this would quickly place a burden on the reader and make your work difficult to read. In the PhD, this might result in an unpleasant supervisory reading. In your viva, it could result in needless questions, where the reader has had difficulty following your line of reasoning, and perhaps misinterpreted what you wanted to say. If you are submitting work for publication, then it could similarly result in your peer reviewers misinterpreting what you want to say, extensive corrections, or maybe even a rejection.

So, **rule one: commas indicate dependent clauses**.

Along the same lines, **rule two: commas separate introductory elements**. Students occasionally omit these commas.

- Conversely, using the latter substrate had definite advantages.
- Consequently, it is recommended that this approach should be taken.

Introductory elements seem inconsequential, but they are actually very important. These words guide your reader from one piece of information to the next and allow them to see how they are related, and how they impact on each other. Important as they are, however, they are not part of the main import of the sentence. You can see that we could remove them and still have the sentence make sense:

- Using the latter substrate had definite advantages.
- It is recommended that this approach should be taken.

Much like dependent clauses, introductory elements provide additional information to aid your readers' understanding by providing more context for the information you are delivering. The same rule applies to them as it does to dependent clauses: they should be clearly defined with a comma.

Along the same lines of indicating additional information, **rule three: commas should be used to set off appositives** – nouns which provide extra information on other nouns – off from the rest of the sentence.

- The foremost specialist in this field, **Professor Davis**, will be delivering a talk on Wednesday.
- The noun phrase 'Professor Davis', gives us more information about the other noun phrase 'the foremost specialist'.

Rule four: commas can also punctuate items in a simple list.

We took the following factors into consideration: cost, duration, ease of use, and applicability.

Again, this allows you to see that the comma essentially acts to provide clarity, separating the items in the list. You might also notice that I have placed a comma next to 'and', before the last item in the list. This is known as an Oxford (or 'serial') comma. It can help you avoid ambiguity when the last two items in a list might be perceived as a compound.

If you do decide to use the Oxford comma, then it's best to be consistent about it, and apply the same style throughout your writing.

Rule five: commas separate *coordinate* adjectives.

If English is your first language, then you may not be aware that it has a very specific hierarchy when employing adjectives. Adjective word order is as follows:

> Opinion, size, physical quality, shape, age, colour, origin, material, type, purpose

Some examples might help here from each category:

Opinion	Difficult, novel, notable, appealing, obscure
Size	Large, miniscule, small
Physical quality	Smooth, rough
Shape	Round, hexagonal
Age	Young, old
Colour	Blue, green, red
Origin	British, French, Japanese
Material	Metal, silicon, carbide
Type	L-shaped, specialised, general-use
Purpose	Probing, measuring, weighing

Native speakers implicitly know this order, and it sounds very wrong if the order is confused. Compare the following:

- A new silicon container was used.
- A silicon new container was used.

The second example sounds jarring, although you may not previously have been aware of why it sounds wrong.

The rule regarding the use of commas with adjectives is that if we use two adjectives from the same category, then we have used *coordinate* adjectives, and a comma is required. For example:

> Heat stress has a significant impact on the productivity and yield of crops grown in the **hot, arid** zones of the world (Ramsay, 2014, p. i.).

If we use adjectives from two different categories, then a comma is not required, as in the example offered below.

> A new silicon container was used.

These aren't the only five rules about commas, but they're perhaps five of the most important ones when it comes to maintaining clarity using the most common functions for someone in your position.

Common comma errors

We have already discussed the damage that can be caused by omitted commas in a lengthy sentence. The next most common types of comma error are as follows:

The comma splice

> The changes were made, the results were recorded.

A comma splice is **when a comma is used on its own to try and link two independent clauses**. While the comma is, as we said, a versatile piece of punctuation, it is not strong enough to link two independent clauses. If you want to use it in this way, you must use a conjunction (a linking word) alongside it.

> The changes were made, and the results were recorded.

If you want to use a conjunction with a comma in this way, then you use a very specific conjunction known as a 'coordinating conjunction'. These are conjunctions which join clauses of equal importance (note the same terminology we introduced to you in the previous section, when we talked about two adjectives that belong to the same category: 'coordinate', meaning 'jointly/equivalently ordered [in a hierarchy]'). The coordinating conjunctions are: for, and, nor, but, or, yet, so.

The next common type of error is to have **a comma separating the subject and the verb**:

> Crops grown in eastern regions of Africa using XX fertilisers developed since 1988, often demonstrate this abnormality in their growth patterns.

We have a very long subject in this sentence:

> Crops grown in eastern Africa using XX fertilisers developed since 1988

The student has read over this sentence, and they have decided that it should have some form of punctuation, since it seems long and complex. Unsure of where the punctuation should be placed, they've inserted a comma at a point where it seems like the sentence changes focus. However, if we read the clauses on either side of the comma, we can see that neither will stand alone as an independent clause.

- Crops grown in eastern Africa using XX fertilisers developed since 1988
- Often demonstrate this abnormality in their growth patterns

You might see sentences where the subject and verb are separated by a dependent clause.

- This new line of enquiry, first discussed at length in Smith et al's pivotal 2013 paper, will be examined here in more detail.

Here, though, the subject and verb are still technically part of the same sentence. The commas demarcate the subordinate clause, which can be removed (one of the defining features of a subordinate clause). Removing it leaves us with:

> This new line of enquiry will be examined here in more detail.

You can question the writer's stylistic choice in this example – placing long subordinate clauses in the middle of sentences is detrimental to readability, but it is not a subject/verb separation error as we saw in the original African crop sentence.

Semi-colons

The most common student writing error seen when using semi-colons is general confusion over what their purpose is, which leads to them being used inappropriately when a comma should be used instead. This is often the case when punctuation has been taught as a series of 'pauses', with a comma as the shortest pause, a semi-colon as a slightly longer pause, and a full stop as the longest pause.

It's much more useful to simply know the two situations in which a semi-colon is appropriate, which will allow you to exploit them for maximum effect in your writing.

The first use of the semi-colon is to separate clauses in a compound sentence, as we mentioned when discussing the examples on p. 107.

- Some researchers work well alone; others work well in a team.
- The first set of results seemed especially promising; the second set presented significant problems.
- Such responses to drought may include overt phenotypes like growth arrest; this is advantageous as the cessation of growth means that any metabolic activity can be focused on avoiding or ameliorating the effects of the stress (Ramsay, 2014, p. 23).

The comma, as you have seen, is a versatile piece of punctuation which allows you to effectively sub-divide information within sentences and indicate to your reader which is of primary importance, and which is supporting material. As we discussed, though, the comma is not strong enough to link two independent clauses on its own: it always requires a conjunction. A semi-colon, on the other hand, can be used *without* a conjunction.

A semi-colon is also used to provide clarity in complex lists. For example:

There were several factors taken into consideration: cost, time and availability; impact, interest and publicity; and relevance, benefits, and interdisciplinary reach.

We could punctuate this list using commas alone.

> There were several factors taken into consideration: cost, time, availability, impact, interest, publicity, relevance, benefits, and interdisciplinary reach.

However, it makes life easier for our reader if we can find some way to thematically sub-divide the list, in this example, by practical concerns, external interest and success. When we choose to sub-divide a list in this way, semi-colons are used to separate categories within the list.

Colons

Colons are rarely used in the writing that we see from students, possibly because they're not sure of how they should be used, and tend to avoid them. Colons only have two jobs, however, and can be used effectively in your writing once you are aware of what they do.

Colons punctuate the explanation or amplification of a statement

> There was one common finding across all studies carried out in the last decade: raised levels of magnesium in the bloodstream.

You can see that when it is used in this way, the colon can really draw attention to the second half of the sentence. Using a colon to announce information in this way tells the reader that it is important, and that they should pay particular attention to it. Along with simple sentences, then, colons are a useful tool for when you want to make completely sure that a key piece of information is highlighted within the text.

Along the same lines, **colons also introduce a formal list**. We saw this in the semi-colon example.

> There were several factors taken into consideration in designing the project: cost, time, availability, impact, interest, publicity, scope, benefits, and interdisciplinarity.

Colons do not have to be used to introduce every list. A short, informal list does not require a colon to introduce it. The list in the example above, however, seems important. The researcher is describing the deciding factors that were taken into consideration when designing their project. It's likely that they want to make sure the reader pays attention to this information, and so they've decided to introduce it with a colon.

We hope that the information offered here doesn't seem like a set of rules designed to constrain you and slow your writing down, but that it instead empowers you, by giving you control over exactly how your reader understands and interprets your work.

We don't suggest that you attempt to write your first drafts with all this information in mind. This would be likely to hinder rather than help you. What we would suggest, however, is that you think carefully about it when you're developing your product draft. Does the sentence structure you've used best serve the point you're trying to make? Have you buried important information in a lengthy multi-clause sentence?

Equally, sentence structure can be used to help you analyse content. Carefully read any frequent complex and compound/complex sentences you might have. Is all of that supporting information necessary? Could you be more judicious? On the other hand, too many simple sentences could point to a failure to provide your reader with adequate context, or a lack of awareness of nuance and appropriate tentativeness. An understanding of sentence structure has multiple uses in your writing process.

Our overview of punctuation is equally offered with the same aim in mind: to give you complete control over how you deliver your work by understanding how you can use different punctuation marks to guide your reader through complex subject matter that requires a high level of concentration.

The following summary covers the material discussed in this chapter. It should be useful as a quick reference when you are editing and proofreading your work.

Summary of sentence structure and punctuation

Sentence structure
- Simple sentences: one independent clause.
- Compound sentences: two independent clauses.
- Complex sentences: one independent clause and at least one dependent clause.
- Compound/complex sentences: two independent clauses and at least one dependent clause.

Commas
- Punctuate main linked clauses (when used with a conjunction).
- Indicate dependent/subordinate clauses.
- Punctuate introductory elements.
- Punctuate appositives.
- Punctuate coordinate adjectives.
- Separate items in a list.

Semi-colons
- Punctuate main linked clauses.
- Separate categories in a complex list.

Colons
- Introduce the explanation or amplification of a preceding statement.
- Introduce a formal list.

CHAPTER **10** **Paragraphs**

We see students' chapter drafts at every stage of their work, and paragraphing is almost always an issue. Dealing with large, complex topics in detail tends to naturally lead to very long, dense paragraphs. Within these paragraphs, students frequently struggle to discern where the topic has shifted and when a new paragraph should be taken, because they're acutely aware of how the issues they are discussing are all closely inter-related.

This can be especially problematic for students who do not have English as a first language, and whose first language has conventions that place the burden of interpretation on the reader, as opposed to the writer. They frequently receive feedback regarding 'difficult to read' paragraphs, but have no sense of what makes the paragraph difficult to read, or how they should begin to address the problem.

In this chapter, we'll discuss a variety of paragraph models that can be used in order to ensure that information is conveyed coherently and concisely. We'll explain how effective paragraphing can guide your reader through even very dense, specialist material, and how a knowledge of paragraphing techniques can help make writing seem less overwhelming by enabling the work to be broken down into small, manageable chunks.

Addressing paragraphing problems can produce a dramatic improvement in writing in terms of both clarity and structure. We hope that this chapter will enable you to effect these changes in your own work.

In Chapter 9, which discussed sentence structure, we said that sentences should make one main point. This is equally true of paragraphs, which should essentially function as one coherent unit within your chapter. Every paragraph should be:

- focused/unified – the paragraph should be focused on one point/idea
- fully developed – the paragraph should fully cover this one point/discuss this idea with supporting data/evidence/examples where appropriate

- coherent – the paragraph should progress logically, without wandering off-topic, or attempting to cover too much material.

No matter what 'type' of paragraph you're writing (whether it describes a process, or compares and contrasts two ideas, or presents a theory), it still must adhere to these principles in order to be a successful paragraph. Your reader expects to be able to:

- understand how this paragraph is relevant to the rest of the chapter
- see from the first sentence what it will cover
- be guided through the information it contains
- be left with a sense of how the next paragraph will logically follow on from this one.

We will now describe the four main types of paragraph you might use. Bear in mind that you can use your own judgement in terms of modifying these structures to effectively communicate your point. The key is to remember the principles discussed above, and to keep your readers' needs uppermost in terms of clarity and coherence.

Types of paragraph

Argument
This type of paragraph follows a very specific structure.

> 'Writing for publication has several benefits for doctoral students. One particularly positive impact it can have is on performance in the viva. Out of 500 researchers who were surveyed, 77% "strongly agreed" that they felt publication made them feel more confident in the viva. There could be several reasons for this, from a sense that their work had already been recognised as novel and valid in their field, to resilience built by going through the peer review process. It should be pointed out, however, that while this would appear to be a significant benefit in the eyes of the student, some supervisors still express concerns that focusing on publication can detract from progress in the PhD.'

The paragraph has a topic sentence – 'Writing for publication has several benefits for doctoral students'. It goes on to provide specificity via an elaborating sentence, which talks specifically about performance in the viva. The reader, then, is made fully aware of what will be covered in the paragraph.

We next have a piece of data, and then some explanation of it. The order here could change, you could make a statement and then back it up with supporting evidence, but both elements have to be present.

Lastly, we have a concluding sentence that sums up the paragraph, and gives some indication as to next directions.

Certain features of this paragraph type can also be found in the sentences that follow, particularly the topic and elaborating sentence. Look out for them in each paragraph to get a sense of how they set context and apply parameters for the material that follows.

Process
This type of paragraph is built through the description of a process.

'Serotonin (5-hydroxytriptamine) is a neurotransmitter. It plays a role in regulating the appetite (Bear et al., 2016). While dopamine is released in order to stimulate appetite, serotonin is released later in the process, when food is being consumed. This activates the 5-HT2C receptor on cells that make dopamine, which has the consequence of preventing more dopamine from being released. This means that serotonin stops the appetite. This is only one role serotonin plays in the nervous system.' (Adapted from Bear et al., 2016)

You can see that this type of paragraph, even though it describes a complex process, is unified and coherent. It has a strong topic sentence – 'Serotonin (5-hydroxytriptamine) is a neurotransmitter' – which sets parameters for the focus of the information that follows. The elaborating sentence provides additional focus by specifying which particular role will be discussed throughout the rest of the paragraph.

Contrast/Comparison

'Enantiomers and diastereomers are both types of stereoisomers. They share the same molecular formula. They also share the same type of connectivity (Fox and Whitesell, 2004). However, enantiomers are non-superimposable mirror images of each other, diastereomers are not. This means that while enantiomers share the same physical and chemical properties, diastereomers do not, and can have different physical and chemical properties. Consequently, these different properties mean that diastereomers can be more readily purified than enantiomers.' (Adapted from Fox and Whitesell, 2004)

Again, while this is a different type of paragraph than the paragraph by process, you can see that it fulfils our four guiding principles for a good paragraph. The reader expectations we discussed are also met. They have a clear sense of what the paragraph will cover from the clear topic sentence. They are clearly guided throughout, and the paragraph concludes with an indication of what will be discussed next (the purification of diastereomers).

Detail

'The hippocampus is located in the medial temporal lobe of the brain, at the edge of the cerebral cortex (Watson et al., 2010). It can be further sub-divided into the cornu ammonis and dentate gyrus. In turn, the dentate gyrus can be further broken down to the hilus and fascia dentata. It is often described as being shaped like a seahorse, hence the name – hippocampus – from the Greek hippos + kampos.' (Adapted from Watson et al., 2010)

The focus on describing one object means that a paragraph built by detail is likely to be appropriately coherent and unified. This paragraph type is only likely to encounter problems if the object being described is too large and complex, which could result in the provision of too much detail. Here, for example, trying to build a paragraph on the entire medial temporal lobe of the brain might have caused problems.

As mentioned in the introduction to this chapter, however, we find the students' paragraphs can often wander from the three principles we discussed, frustrating their readers' expectations and causing comprehension issues. We'll look in more detail now at the most common problems we encounter.

Common paragraph problems

The topic sentence is too broad

Read the following paragraph.

> 'Attempting to understand how the brain deals with music is complex. The frontal parts of the brain seem to be involved when music elicits an emotional response (Warren, 2008). The brains of musicians and non-musicians seem to deal with music input differently. Parts of the brain which deal with issues such as motor tasks seem to exhibit structural changes. There are similarities in how the brain deals with music and how it deals with language. For example, Broca's area is implicated in both. Rhythm and tonality both involve complex processes crossing several different structures. When the rhythm is particularly complex, even more parts of the brain become involved, such as the cerebellum. Damage to the amygdala can affect the emotional perception of music, according to a study carried out by Gosselin, Peretz, Johnsen and Adolphs in 2007. Brain damage can impact on musical ability, depending on the part of the brain which is affected, and the role it performs in terms of music ability.' (Adapted from Warren, 2008)

It is easy to see here that the topic sentence introduces a broad, high-level concept, but that is not what follows. The author has tried to use it as an introduction to a very wide range of concepts about the brain, while simultaneously describing these at a very fine level of detail. It's not incoherent: the material presented is all connected to the central topic, but too many important points are made. This makes life difficult for your reader, who is trying to follow, understand and retain this information. There's also the risk that any original points made (e.g. comparing music and language) might be lost, because they're buried in too much information.

Students can often find it difficult to remedy paragraphs like this when they encounter them. Everything contained is relevant, after all, and cannot simply be removed. How might you begin to tackle a paragraph like this?

Since the root of the problem lies in the breadth of the topic sentence, this is the place to start.

- Attempting to understand how the brain deals with music is complex

There are several points that could branch off from this statement, which is why the paragraph is so unwieldy. These can be identified as follows:

- Brain structures involved in processing music.
- Structural differences between the brains of musicians and non-musicians.
- Similarities between the processing of language and music.
- The impact of brain damage in music ability.
- The specific impact of damage sustained by the amygdala.

If we see each bullet point as a possible topic sentence, instead of an element within one overloaded and complicated paragraph, we now have a potential plan for five paragraphs that will give you the opportunity to fully expand on each topic. This enables you to present an appropriate level of supporting evidence within each paragraph, as well as ensuring that any important observations can be foregrounded.

The topic sentence is missing

Alternatively, paragraphs missing a topic sentence have a different kind of problem. The information that's contained within the paragraph might all be unified, coherent and fully developed, but without an appropriate topic sentence, it's not immediately apparent to the reader how they ought to contextualise this paragraph's information within the wider chapter.

Consider the impact it has on one of the paragraphs we've already seen.

> 'The hippocampus can be subdivided into the cornu ammonis and dentate gyrus. In turn, the dentate gyrus can be further broken down to the hilus and fascia dentata. It is often described as being shaped like a seahorse, hence the name – hippocampus – from the Greek hippos + kampos.' (Adapted from Watson et al., 2010)

Do you see the effect that removing context and proceeding immediately with detail has had? The solution here is obvious enough: ask yourself how this paragraph relates to the chapter, and what context any reader who is less familiar with the content than you are might need. It's possible that the previous paragraph could have laid out the main structure of the hippocampus, in which case this paragraph would serve as elaboration on that point, but the fact that this paragraph ends with a very top-level piece of trivia would then seem out of place. Instead, replacing the topic sentence that we removed would fix the problem (see the paragraph on p. 128).

Lack of interpretation

A common problem in PhD students' paragraphs is a lack of interpretation or explanation of the data they've presented. It's easy to see how this might happen. You work on this project all the time. No one is more familiar with this data and what it means than you. As such, however, it's important to remember that everyone else will require explanation in order both to understand the data, and why you have chosen to present it in the paragraph. Read this example again:

> 'Writing for publication has several benefits for doctoral students. One particularly positive impact it can have is on performance in the viva. **Out of 500 researchers who were surveyed, 77% "strongly agreed" that they felt publication made them feel more confident in the viva.** It should be pointed out, however, that while this would appear to be a significant benefit in the eyes of the student, some supervisors still express concerns that focusing on publication can detract from progress in the PhD.'

The risk here is that without your guidance on how that statistic should be interpreted, you're leaving your reader to draw their own conclusions, which could lead to a misinterpretation of the point you were attempting to make.

You might find that you produce a paragraph like this in your early process drafts. This is fine in the process draft; just remember that when you write the *product* draft later, you should be focusing on the reader, and they are likely to require an explanation.

The paragraph contains too much material

Alternatively, your paragraphs might be too bloated in the middle. This is often the result of a lack of judiciousness in the presentation of data or supporting information. You may also have provided too much explanation. Let's use the argument paragraph as an example again.

> 'Writing for publication has several benefits for doctoral students. One particularly positive impact it can have is on performance in the viva. Out of 500 researchers who were surveyed 77% "strongly agreed" that they felt publication made them feel more confident in the viva. There could be several reasons for this from a sense that their work had already been recognised as novel and valid in their field to resilience built by going through the peer review process. 27% of researchers expressed a desire for more training to prepare them for the viva with a particular focus on some kind of online training package. It should be pointed out, however, that while this increased confidence would appear to be a significant benefit in the eyes of the student some supervisors still express concerns that focusing on publication can detract from progress in the PhD.'

The impact of the additional statistics can be easily seen. They are related to the topic of the viva, which is how they've crept into this paragraph, but they're not really related to the topic sentence. As such, the paragraph abruptly loses focus, and starts to feel cluttered. Further, these statistics, which might well be relevant to the overall discussion, don't get room for their own examination, but simply get lost in the middle of the paragraph.

The reasons behind this paragraphing problem are not necessarily negative. If you have a lot of interesting data that you're excited about, then you might find it difficult to be selective in what you present per paragraph. If you're keen to ensure that the reader has fully understood something, then it's tempting to restate the same point in a couple of different ways, just to be sure that they've definitely absorbed what you wanted to say. The end result, however, is still a paragraph that would benefit from some pruning in order to retain a clear focus.

The paragraph doesn't have enough material

This is reasonably common. It tends not to come from a student deliberately choosing to write a very short paragraph. Instead, it tends to be the student's response to finding an overly long paragraph that requires editing. Instead of thinking carefully about the elements we have mentioned (the breadth of the topic sentence, etc.), the student instead takes material from the end of the paragraph and creates a very short paragraph immediately following it. For example:

> 'One of the reasons that smell memory and emotion seem to be connected has to do with the structure of the brain (Wilson and Stevenson, 2003). The amygdala which is involved in processing memories to do with emotions such as fear and anger plays an important role. It is located close to the primary olfactory complex which deals with olfaction. The amygdala has direct connections to the hippocampus which deals with short and long-term memory. It can be seen to be important for survival reasons that mammals develop and store memories to do with fear and anger. The primary olfactory complex amygdala and hippocampus are all part of the limbic system although the concept of the limbic system is contested by some neuroscientists (Heimer et al., 2007).'
>
> (Adapted from Heimer et al., 2007; William and Stevenson, 2003)

There's no real reason for this section to have become detached from the rest of the paragraph other than, as we pointed out, the author's concerns over length. As it stands, it could easily be reintegrated with the rest of the paragraph.

Connectives

Dense paragraphs can be very difficult to follow. Sometimes, however, they're unavoidable. You are writing complex material with a highly specialised vocabulary, and you might simply have to deliver a particularly detailed chunk of information. How, then, can you help your reader in this instance?

The key here is the use of connectives. You might recall that we first mentioned these in Chapter 2. Connectives are words that provide clear links between pieces of text, letting the reader fully understand how they relate to one another, and how they should interpret the information. They're invaluable if you have to, for example, talk your reader through an especially complicated process.

This table lists some typical examples, as well as the role they can play in advancing your paragraph.

Causal	Consequently, as a result, because, thusly, since
Temporal	Previously, when, next, simultaneously, presently, while
Contrasting	Alternatively, alternately, in contrast, in comparison, conversely
Emphasis	Crucially, vitally, particularly, especially, most importantly
Summarising	In essence, essentially, in summary, in brief, in short, basically
Additional	Moreover, additionally, also, first, second, third, furthermore

Let's read our contrast/comparison paragraph without connectives to get a sense of the importance of the role they play:

> 'Enantiomers and diastereomers are both types of stereoisomers. They share the same molecular formula. They also share the same type of connectivity (Fox and Whitesell, 2004). Enantiomers are non-superimposable mirror images of each other. Diastereomers do not share this chirality. Enantiomers share the same physical and chemical properties, diastereomers do not, and can have different physical and chemical properties. These different properties mean that diastereomers can be more readily purified than enantiomers.'

We've only removed three connectives, but you can see the impact it's had on the paragraph. The reader is offered no help in understanding how these pieces of information relate to each other. The paragraph also seems very staccato. The information is all related, but the writing style doesn't help to convince the reader of that.

A good exercise to practise using connectives effectively is to read over paragraphs you've already written and count how many connectives you find within each one. We often ask students to write a paragraph during one of our workshops and then ask how many connectives they've used. With very few exceptions, the answer is usually one or none. This isn't to say that every single sentence needs a connective to open it, but you should try to consider the paragraph from the reader's perspective and ask where support might be needed to bridge sentences.

You could also read published articles and theses. If there's a section of text that you find particularly difficult to follow, ask yourself if adding connectives would aid your understanding.

As we stated at the beginning of the chapter, we've found that paragraphing can often present difficulties for the students we meet – unnecessarily inhibiting their clarity and causing structural problems. The techniques we have described here give you the tools you need to give your work a strong structure that enables it to convey complex information clearly and concisely.

CHAPTER **11** Editing and Proofreading

Good editing and good proofreading are vital in the production of a tightly written and comprehensive piece of work. They are also demanding tasks which require a great deal of concentration from the writer, an objective overview of large sections of text, and an eye for fine detail. We find that students can often fall into unproductive habits in editing and proofreading: endlessly tweaking early drafts, and/or leaving the bulk of the task until the last minute, both of which can leave them feeling overwhelmed by the task and unlikely to do good work.

This chapter will discuss the editing and proofreading processes. We will emphasise here the idea that editing should be an ongoing process throughout the writing of the thesis, which reflects your ability to refine your thinking and communicate your ideas clearly to the examiner.

As we discussed in previous chapters, our ultimate aim is to give you control over your writing. You'll find that the practical tips and techniques in this chapter can be put to use by analysing published journal articles and theses in order to hone your skills for editing your own work, where it can initially be difficult to have the objectivity required to spot common problems.

We will talk first about the differences between editing and proofreading, two terms that we find students often use interchangeably. Once we have established the differences between the two, we'll talk about the techniques you can employ to tackle both tasks in order to make them both manageable, and to ensure that time you spend on them is effective. We will also employ some of these techniques on sample texts, to give you an idea of how effective they can be, and how you might begin to implement them on your own work.

The difference between editing and proofreading

A simple definition is that while **proofreading** looks at the nuts and bolts of the text, such as punctuation, referencing, spelling and formatting, **editing** is about the deeper factors underlying how you've communicated, such as argumentation, structure and clarity.

Let's examine both individually in more detail.

How to edit effectively

Editing is a deeper and more comprehensive process. Editing is not about the superficial details; editing is about reading for **sense** and **coherence**.

- Is there a logical flow *within* paragraphs?
- Is there a logical flow *between* paragraphs?
- Have you backed up your points with sufficient and appropriate supporting evidence?
- Have you been judicious in your use of that evidence, or are your paragraph structures staggering under the weight of too much material?
- Does the overall structure of the thesis 'fit' the content?

What about **clarity**?

- Have you clearly stated your intent, what the chapter will do and how it relates back to the wider piece of work?
- Is your word choice precise?
- Is your phrasing concise?
- Have you clearly defined any terms which may be unfamiliar to readers outside of your exact research niche, or which need to be used with particular precision in your work?

Check your tone.

- Is your word choice appropriate?
- Have you avoided the use of phrasal verbs and other more informal phrasing (e.g. 'talk about' vs. 'discuss', or 'look at' vs. 'examine')?

Reverse outlining

A particularly powerful editing technique that we often use when helping students analyse their writing, and that you might find particularly useful when editing chapters, is reverse outlining.

In Chapter 9, which dealt with paragraphing, we talked at length about the importance of an effective topic sentence. Topic sentences give the reader the controlling idea of the paragraph, letting them know exactly what it will cover, and sometimes also relate back to the overall topic of the chapter.

Reverse outlining uses these topic sentences to provide a skeleton of the chapter's structure, allowing you to see where any structural problems might have occurred, while avoiding a complete read-through of the material, which can lead to getting lost in the content when you're trying to focus on structure.

The steps involved in carrying out an effective reverse outline are as follows:

1. Open a new blank document.
2. Pull up the section of your thesis you want to edit alongside it.
3. Copy and paste the first sentence of every paragraph into the new document.
4. Change these sentences in the new document into a numbered list.
5. Read through the list.

You should be looking for three things in the reverse outline:

1. Does the structure flow logically? When you read it through from beginning to end, is there a sense of progression?
2. Are there any **gaps**?
3. Is there any **repetition**?

Dealing with gaps

If there is a gap between two of the sentences, but you think that there is still a sense of progression there, then it's possible that a bridging paragraph is needed. Before you write this, though, carefully read the content of the first of the two paragraphs, before the gap. It's often the case that when there's a gap like this, the paragraph that precedes it has tried to cover too much material, and the necessary bridging paragraph is already there, but buried. Read the paragraph carefully. Does the focus shift? Is the topic sentence slightly too broad?

Dealing with repetition

There are two main reasons that repetition tends to occur.

Instead of starting a fresh draft, as recommended in Chapter 2, **you have cobbled together a Frankenstein's monster draft**, consisting of paragraphs from several different iterations of the same draft. This can often result in the accumulation of paragraphs that are similar enough that they seem repetitive when placed alongside each, but are slightly different in terms of phrasing, or tweaks in terms of what is covered.

It can be difficult to bring yourself to excise paragraphs like this, as they each might, in a small way, contain something of value.

If you find it hard to delete work, open a new document and paste the repetitive paragraphs into it. Read them carefully, and write a one-sentence summary beside each. If the topic sentences were truly repetitive, then the summaries are also likely to either be the same, or virtually the same. If you have three paragraphs that are effectively making the same point, then you'll need to write an entirely new paragraph that can be reinserted into your draft.

The other instance in which repetition might occur is if **you are very keen to ensure that the reader understands a specific point you have made**. When this happens, it's common to reiterate the point, attempting to phrase it slightly differently each time in order to ensure that the reader has taken the information in.

Again, follow the same process. Copy and paste the paragraphs into a new document. Write a one-sentence summary of each as a check to see exactly what each one covers. If they all say the same thing, then you need to choose which says it best, or write a new paragraph.

After you have addressed gaps and repetition, make a new reverse outline. This one should flow more effectively, reflecting the improved structure of the chapter.

A quick tip – you can also flip reverse outlining and use it to *build* chapters. If you can construct a bullet-pointed list of what you would like a chapter to cover, then you can build a paragraph around each of these stand-in topic sentences. This can be especially effective if you're having difficulty getting started, or if you find that you'd prefer your process draft to have a little more direction.

How to proofread effectively

Proofreading, as we have said, involves reading for superficial errors. These include grammatical errors, such as misplaced punctuation. It also involves searching for spelling mistakes, formatting problems and referencing errors.

Although proofreading deals with superficial errors, it is still hugely important. Your attention to detail in your work is a sign of professionalism. Work that contains errors, no matter how small, can appear sloppy. It can also plant a seed of doubt in your readers' minds about where else you might have made careless mistakes, and failed to pay attention to details.

Common problems

Although the actual process of proofreading might seem simplistic, in that it only deals with surface errors, it can actually be quite a difficult task to carry out. There's a few reasons for this.

- You're invested in the content, and it's hard to read the document without becoming distracted by it.
- It requires a great deal of attention to detail, which you can only sustain for a finite period of time.
- It's not particularly exciting work, and it's easy to become bored and distracted, leading you to miss errors.

Tips and techniques

This is a good place for you to use the kind of reflective techniques we discussed back in Chapter 2. Try to proofread in blocks of time. You could start with something small, like a 15-minute block. If your attention starts to waver at any point, take note. Try this method for four blocks and see if your distraction point seems to be consistent. Double-check the proofreading you carried out after that point each time. Have you missed any errors in the text? This should give you a reasonable idea of how long you can proofread for and still be effective.

You might be able to sustain your concentration for a longer period of time if you adopt another technique: only searching for one specific type of error in any one session.

As we said previously, proofreading demands that a great deal of attention is expended on small details. Dividing your efforts by reading for punctuation errors *and* formatting problems, for example, places a heavier demand on your attention. Selecting only one kind of error to search for in each session means that you'll be able to sustain your focus for a longer period of time.

Following on from this, if you're searching for errors with a specific type of punctuation mark, then you might find it helpful for focusing your attention on them to highlight or circle *every* instance of that punctuation mark in the document you're working on.

You should learn what your most common mistakes are, and to be watchful for these when you're proofreading. No one makes every kind of mistake all the time. We tend to have habitual errors, and if you know what mistakes you commonly make, then you'll find it easier to spot them. Pay attention to the mistakes flagged by whichever word processing software you use. Take note of which ones seem to be most frequent. Your supervisor might also be able to give you feedback on this, if you ask them what mistakes tend to crop up most frequently in your work.

Reading the work backwards is another technique you could try. You can do this word by word if you're searching for typos and formatting problems, or sentence by sentence if you're paying attention to grammatical problems. This technique is designed to prevent you from reading for larger content issues, since reading for content moves more into editing territory. Reading backwards can be time-consuming, but those who can adopt this method say they do find it effective.

Word clouds offer a different way to visualise your written work, and can be a very effective and simple way to spot excessive repetition. A word cloud is a visual representation of a section of text. Words are represented according to frequency, with those used most often largest.

Ideally, you would hope that subject-specific words and terminology relevant to your project would be biggest, but you might also spot an over-reliance on certain words and phrases, which you can then address when redrafting.

General tips for editing and proofreading

- Read your work aloud. You'll be surprised at how many more errors you spot this way, as well as larger issues, such as a lack of flow between paragraphs. We occasionally have students referred to us for individual appointments by supervisors who have concerns about their English. In the vast majority of cases, these students pick up on errors as soon as they hear them, despite having left them in the document.
- When you read your work aloud, don't cheat by adding or removing punctuation when you spot problems. Read the work exactly as it is written. Record yourself if you really want to get a sense of the flow of the overall chapter.
- You might find that your proofreading and editing is more effective if you print your work out and work from the physical page. We both find it very difficult to edit and proofread effectively from the screen, and always have to print work out in order to spot problems. (Bonus benefit: a stack of printed paper also gives a realistic sense of how much of your thesis is written so far, which you just can't replicate on a single glass screen.)
- Some people find that editing and proofreading require even more attention than writing the drafts. Although the boot camps we offer at our institution are designed for generative writing, that is, simply getting words on the page, we've increasingly found that people like to use the time for editing instead, finding that both the sustained focus and silence enable them to do the job effectively.
- Some students have also reported that they like to print their work off in a different font when they're proofreading and editing, and that this gives them a sense of reading the work with fresh eyes.
- Try to give yourself a reasonable gap between finishing your draft and starting the editing and proofreading processes. When you've been working closely on a draft for a long period of time, it's easy to stop seeing mistakes, and also to lack the appropriate distance to be objective about the work.

- If you find it hard to delete work, then instead of doing this, place excised sections of text into a new document which you then save and keep. Larger sections of text, such as a discarded chapter, might even be able to form the foundation of a conference paper or a publication. Remember, the shape of the thesis will shift as your research develops and takes new directions. Sometimes, that means making difficult decisions.

Insights from a researcher

Don't be afraid to get rid of sections that aren't working. My thesis was eight chapters long, with three findings chapters. However, those three chapters were finalised after about five or six attempts at making sense of the findings, and I didn't hit on the final orientation until the final six months of the write up (I was part-time, so for full-time, say two to three months). And then there was the occasion when I deleted 34,000 words because they just weren't working for me, and trying to edit was just making things worse.

CHAPTER **12** # Making the Document Look Like a Thesis

We've advised you elsewhere in this book that you should start writing for your thesis as early as possible. At some point, however, there will probably come a time when you think of yourself as no longer working mainly on the research phase, and shift primarily into the writing phase. At this stage, you will probably start thinking about what the final document will actually look like.

Adjusting elements of formatting and layout is an extremely time-consuming process. Most students we spoke to estimated that it took at least four or five times longer than they originally budgeted for. Of course, you'll submit your thesis to your examiners and you'll be told at your viva to make corrections before handing in your finalised version, but don't think of this as a chance to put off improving your formatting until that very last stage. The main readers of your thesis – perhaps the only readers, ever – will be your examination panel. If you're going to make the document well-formatted, do it for them.

There are two good reasons to do so. The obvious one you might think of is so that you can take pride in what you submit. A professional-looking thesis will contribute to showing the examiners that you're a careful, methodical, professional academic. However, one of our principal objectives in writing this book has been getting you to think about the experience your readers will have. With that in mind, think about how much more easily your examiners will find their job if you make things uncomplicated for them.

If you do this, **they'll be in a better position to focus on assessing your competence as a scientist**, as they won't have to battle through impenetrable layouts, poor cross-referencing or badly organised data.

In this chapter, we'll outline the things you should pay attention to by drawing your attention to why they're particularly important to your reader, or why they're particularly challenging and therefore likely to require more work than you might expect.

There are lots of (probably quite hidden) tricks that your thesis-writing software has to help manage the task. Without going into the specifics of any one particular tool, we'll try to point out what functions you should be looking for.

Making life easier for your reader

Your reader needs to be able to:

1. Quickly navigate through the document.
2. Follow your cross-references.
3. Easily locate figures and tables.
4. Easily comprehend the contents of figures and tables.
5. See how each section of your thesis relates to the others.
6. Read and understand the text (for more on this, see Chapter 11).

Technical challenges for you include:

1. Working with a large file, which means:
   - lots of scrolling in your editing software
   - challenges of computer processing power, especially if you need your computer to render lots of images in a long document
   - managing the significant risk of losing lots of work should your file(s) be lost or otherwise corrupted
   - having consistency in your formatting choices (e.g. headings, fonts, spacings and other design elements).

2. Adhering to implicit professional standards (i.e. giving your thesis an overall 'polished' look – can you imagine a published article using the same style choices you've made?)
3. Adhering to specific formatting guidelines from your own institution.

Cross-referencing

Your thesis will be large and you'll regularly need to make cross-references between sections. For example, when you mention a method during your results chapter, you don't need to write out the entire protocol; instead, you'll tell your reader you did something 'according to the standard protocol (see Section 2.3.8) for a duration of 45 minutes'. Your numbering system therefore needs to be rigorously observed. If you're off by even just one section, it might seem completely obvious and intuitive to you that the relevant section is just a few paragraphs down the same page, but the immediate effect on your reader will be that you were confused, your thesis is messy and that they can't automatically trust any of your other cross-references. If they sense that you haven't paid attention when the stakes are as high as they are with your thesis, will they trust that you've been methodical when conducting your day-to-day research?

Navigating the document

Your numbering system should reflect the hierarchy of your thesis. 'Section 1' or 'Chapter 1' should – obviously – be a chapter-level item. 'Section 1.1' should be a sub-heading within that chapter, and 'Section 1.1.1' should be a sub-sub-heading. Don't substitute good, coherent writing with smaller and smaller section divisions. If you find yourself with too many sub-sections (usually four, i.e. 'Section 1.1.1.1') then consider the ways in which you are making your writing flow together (see Chapter 9 on paragraphing) or consider revising the structure of your draft itself.

Standard document sections

A typical thesis will include:

- Title page – usually involves your name, your thesis title, your institution name (perhaps with logo) and the name of the degree you're submitting it for (PhD)
- Acknowledgements
- Declaration of originality
- Table of contents
- List of figures
- List of tables
- Abbreviations used

- Chapters
- Appendices
- References

Every numbered heading you use should be mentioned here. Investigate your software's features for creating tables of contents; it will save many hours of your life if you don't have to manage the page numbers in the table of contents, list of figures and list of tables yourself.

You'll probably find a similar guide in your university's doctoral training materials, so check to see if there's anything specific to you that we've missed out here.

Placing figures and tables

Figure placement is another important element in making your reader's job more straightforward. You'll have been taught since early in your scientific career that figures and tables should always be referred to in the main body of the text, and that the title, caption and data legend in any figure or table should work together to make that object understandable on its own. Now that you're working on a particularly long document, it's also more important than ever that you place the figures and tables as close *as is practical* to the body text that refers to them.

This usually means at the end of the paragraph. There are exceptions, though.

If the space at the end of a paragraph is very close to the bottom of a page, you might not have space for your figure. In this case, don't just leave a blank space at the bottom. Doing this signals to your reader that they've reached the end of a section or chapter, which will cause them temporary confusion. Instead, continue with the next paragraph of main body text at the bottom of your page, and insert your figure either at the very top of the next page, or wherever you come to the end of that following paragraph.

Unless there are specific, unavoidable reasons that are unique to your particular circumstances, don't insert a table or figure **before** it's referred to in the main body text. This is against the usual convention, which means your reader will assume they've missed the part where you told them to refer to it. If they then break their focus to scan back to find something that doesn't exist, their task as a reader becomes more confusing.

You should also include a list of figures and a list of tables on two separate pages next to your table of contents.

This all has implications for the process of writing your thesis

If you insert a figure or table during your drafting stages, there's a high probability that it will move up or down the page when you edit the text that comes before it. This can become particularly difficult to manage if, for example, the body of a graph is one element on the page, the data legend is another and the caption is floating in a third text box, as these can move independently of each other.

One solution is to temporarily insert page breaks before and after each figure. This isolates the image on its own page, and means that whole page stays as a self-contained unit, moving *en bloc* according to any additions or deletions made to the text preceding it. You should then, of course, remember to either remove these page breaks and properly position your figures on the pages of body text when you prepare your thesis for submission, or be prepared to justify the stylistic choice of dedicating a whole page to each figure if your examiners question you on it.

One quite reasonable justification for doing this might be to allow the examiners space to annotate your figures or tables with any comments they want to raise at your viva. If you go for this option, just be prepared for them to ask you to alter the layout when you submit your corrected thesis for final approval later.

If you are using or adapting (e.g. using only a small portion of) a figure or table from another source, or if you are using someone else's data to create your own custom version in your own style, then you must properly acknowledge that source. See Chapter 8 for more.

Clarity in tables and figures

A well-constructed figure needs three standard elements:

1. figure number
2. title
3. caption

The numbering system for figures should be kept separate from the numbering system for tables. Figures include photographic images as well as computer-generated graphs etc.

Importantly, the titles of your figures will appear in a list of figures at the start of your thesis. For this reason, make sure your titles make sense on their own, and consider whether you need to write them in a standard voice.

Confusing list of figures	Coherent list of figures
Figure 1.1 Measurement of relative strength of forces X and Y	Figure 1.1 Relative strength of forces X and Y
Figure 1.2 Force X is detectable after 45 seconds	Figure 1.2 Detection of force X over time
Figure 1.3 Investigation of force Y	Figure 1.3 Directionality of force Y
Figure 1.4 Force Z	Figure 1.4 Comparison of force Z with forces X and Y

The examples on the left are all doing different jobs: the first describes what was being investigated; the second reports a result; the third is vague; and the fourth misses out the main point, which is that a new variable was added to the list of things under research.

The examples on the right are more consistent: they all describe the aim of the respective experiments.

You can imagine that the examples on the left probably all make sense to a reader who is reading the chapter from start to finish, and might go unnoticed by a student who wrote the titles and didn't edit or proofread for consistency. A reader who tries to navigate directly to a specific part of the thesis by using the list of figures – which sits beside the table of contents, and exists to make this exact task simple – would have a much harder time following the logic and flow of the research.

Captions on your figures and tables are almost always necessary, and they should adequately describe the contents of the figure to an educated reader.

Panelling

Panels may be used to combine several very closely related elements of your results into one master figure. This is helpful when you want to describe related findings very closely together in the main body of your text, perhaps even within the same sentence.

For example, if you were taking photographs of an experiment using four differently coloured filters, you could lay them out in a 2 × 2 square so that the reader could easily compare the intensities of each colour measured from each part of the set-up by virtue of the photos being closely side-by-side. Or, if you were investigating three different physical properties of five new physical materials over time (perhaps to stress-test them before they went to market) you could stack the graphs for each physical property in a panel one above the other. In this way, each of the three properties still gets its own graph, and each of the five materials could still get its own line on each of those graphs, but crucially, the reader could scan along the bottom axis to compare how *all* five materials responded to *all* three types of stress – because the graphs were positioned closely to each other in a panel.

If you choose to do this, think about combining the different graphs/images/etc. in such a way that they fit within an overall square or rectangular shape as this will be neater and easier to fit on the page. Add an overlay with 'A', 'B', 'C', etc. in a consistent place on each individual element. Obviously, these shouldn't obscure any of the data, and if you need to improve contrast between these labels and the contents of the figure, consider adding a solid coloured square to sit behind the letter label.

When you do this, make sure to refer to the contents of each sub-section by letter when you write your caption. For example, this caption (note the accompanying figure number and title) might go with the hypothetical 2 × 2 panel described above:

Figure 2.4 Spectral analysis of unknown object #3 at 400°C

Unknown object #3 was heated to 400°C for 15 minutes in a closed oven. Room lights were turned off, and images were then taken through the reinforced glass door across four spectrographic regions. **A:** At 5 μm (long-wave infrared), most emission was detected from the top of the object. **B:** At 1 μm (short-wave infrared), most emission was detected from the handle of the object. **C:** At 700 nm (red), emission was detected from the handle and also the base of the object. **D:** View of object with normal room illumination, to show overall shape.

Relating sections to each other

Don't be afraid to refer extensively to other chapters and sections. We don't just mean by using cross-references, but rather also by simply reminding the reader what you've already said elsewhere. An examiner might read your thesis from start to finish once, but then they'll very likely be jumping backwards and forwards through the pages as they carry out their critique and think of questions to ask you at the viva. This means they won't always be reading the document as a single, consecutive narrative.

> Phrases such as these can help add meaning to your cross-references by signposting the reader to what kind of information they'll find if they follow them, or by quickly reminding them what they've already read about:
>> 'As was shown in recently published literature (see review in Chapter 1) …'
>>
>> '… and genetic transfer, which will be discussed later in the context of bacterial resistance to antibiotics (see Section 4.5.1) …'

If you find yourself doing this more often than you usually would in a dissertation or a lab report, don't worry. A good rule of thumb is to treat your reader as though they're only going to read the chapter you're currently writing. This is a good analogue of how you, as an experienced reader of scientific articles, will sometimes have chosen to read perhaps only a methodology section, because you only want to find out one specific detail of how to carry out an experimental protocol.

Develop a habit of creating internal signposts like this and your readers will appreciate your efforts.

CHAPTER 13 Writing for Publication

Students are acutely aware that publication is central to an academic career, and can often feel under a great deal of pressure to publish something throughout the course of their studies. However, our experience is that students can often be very underinformed as regarding the process of submitting an article for publication.

In this chapter, **we will explain this process clearly, demystifying publication as much as possible.**

We'll also talk about how to use reviewer feedback, and even rejection, to strengthen your work, and how you can also take this feedback and use it to improve writing in your PhD.

We will also discuss how blogging, podcasting and social media can offer different ways to draw attention to publications, but also offer you an opportunity for reflection on the writing process itself.

Why write for publication?

As we said at the beginning of this chapter, most PhD students are aware of the importance of publication in building a career in research. A strong publications profile is of benefit when applying for postdoctoral positions after the PhD, and there's probably a greater expectation now that PhD students will publish something throughout the course of their studies than there has ever been before.

Transition

We also discussed the PhD as a transition from student to professional researcher. Publishing a piece of work is seen by many as a first entry into the research community, establishing oneself as a professional researcher and part of the scientific community.

Career

Publication is an essential part of the work of a researcher. Familiarising yourself with this process as early as possible is wise to give you a sense of what a career in academia would actually be like. It will also stand you in good stead when applying for postdoctoral positions.

Feedback

Often, our supervisors and examiners are the only people who see our work and offer feedback. Writing for publication means that you'll get fresh perspectives from new people. If your work is published, you'll also get a sense of how it's received by the wider community.

Motivation

The short-term motivation offered by publication can be particularly useful when your PhD deadline seems a long way off.

Confidence

Students we've talked to who published some of their work during their PhD on the advice of their supervisor have said they found the experience rewarding. It was useful to get feedback from people other than their supervisory panel, and they've felt a real boost to their confidence in having their work published – they've felt more like a 'real' researcher. They also say this will make them feel more confident about their viva.

Things to consider

You might want to write for publication, but think first about these potential hurdles.

Lack of knowledge

First of all, if you are unfamiliar with the process, then you might feel stymied by a simple lack of knowledge. This is something we'll address directly in this chapter.

Writing problems

If you're experiencing difficulties in writing drafts for your supervisors, then it logically follows that the idea of writing for a journal might not seem appealing. While writing a journal article is a good way of giving yourself a goal to boost your motivation in the short term, any problems you're experiencing in your writing are unlikely to disappear because you're writing with a different aim in mind.

In this instance, you should read over Chapter 2 ('Establishing a Writing Practice') and apply the principles of reflection and analysis that were described in this chapter to try and identify and address your difficulties.

You might well find that a change of scene – writing to a tighter deadline, to a different style or even simply using a different referencing style – addresses your writing difficulty, especially if it's related to boredom.

Lack of material

If you've simply not yet carried out enough research to warrant a publication, then it's a goal best placed to one side until you have adequate work. Submitting an article that doesn't have enough substantial points to make is likely to result in a rejection, as the work will be deemed too incremental. Do remember, though, that if you have also undertaken a master's degree, then your dissertation from that might be the basis for a publication. It's very likely to be at least near the word limit for most journals, and to deal with one focused point.

Lack of time

If you've talked with your supervisor and feel that there's enough material for a journal article, you then need to think about the practicalities of the process. Again, you should refer back to Chapter 2 and perhaps consider an audit of your time and commitments. Do you have enough time to do this right now? If there are several other commitments looming, then ask yourself honestly if this is something that is a realistic goal for you at this moment in time. Remember, even if the article you have in mind already exists in some form as a draft chapter of your PhD, then it's likely to need a decent amount of revision before it can be submitted, and then require more work following feedback from your peer reviewers.

Institutional expectations and regulations

It is worth pointing out here that while there is an increased recognition that PhD students can benefit from writing for publication, institutions (and departments within institutions) vary widely in the extent to which they encourage or permit this. For example, some institutions are keen to encourage the model of thesis by publication that we mentioned earlier – where publication is obviously a priority throughout. Other institutions might well not have begun to fully engage with this model. Some supervisors might recommend that you aim to produce one paper from your thesis, and then guide you towards achieving this goal. Other supervisors might see publication during the PhD as a distraction and discourage it.

Make sure you talk to your supervisors about your aims, and that you're fully aware of the options open to you.

What makes a good journal article?

A good journal article has a strong story. As well as knowing what it wants to contribute, **it's confident in the value of what it wants to contribute**, and it clearly communicates this to its readers.

Before beginning the process, ask yourself if this is how you feel about your work. **An article that simply records what you did without having a larger sense of contribution is unlikely to be published.** Equally, if you're uncertain or half-hearted about the value of what you've done, this is likely to be reflected in your writing.

Once you have enough material, you're confident about its contribution to the wider academic discussion in your field, and you've discussed things with your supervisor, you're ready to move on to acting on your plans.

Beginning the process

The first thing you need to do is to identify the most suitable journal for your work. Your supervisor is likely to have advice on this, but you should also carry out your own research. Some questions to consider:

- Where do you get most of your own reading material? If there's a particular journal that publishes work relevant to your project, then it's likely that your work might be a good fit for that journal.
- Are you also thinking about impact factor? Impact factor is calculated by looking at the number of times a journal article is cited, thus giving a sense of its impact on the wider field. Most journals have their impact factor clearly stated on their website. It shouldn't be your sole consideration, but it is something for you to think about. You should talk to your supervisor if you need more guidance.
- How frequently is this journal published, and what is the typical turnaround time from the initial submission to eventual acceptance and publication? What are your schedules and deadlines? Do these match up?

Once you've identified a suitable journal, you then need to repeat this process, eventually giving yourself a 'top three'. In the event that your article is not accepted by your first-choice journal, you would then submit to your second choice, and so on.

> Remember, you **must** only submit to **one** journal at a time. You must not send multiple copies of the same article to several different journals.

Make sure you're completely familiar with your chosen journal – everything from its aims and scope to its referencing guidelines. Most journals now exist online, and carry this information prominently, as well as guidelines for authors. For example, here we have the aims and scope for *Nature Cell Biology*:

> *Nature Cell Biology* publishes papers of the highest quality from all areas of cell biology, encouraging those that shed light on the mechanisms underlying fundamental cell biological processes. The journal's scope is broad and specific areas of interest include, but are not limited to:
> Autophagy
> Cancer biology
> Cell adhesion and migration
> Cell cycle and growth
> Cell death
> Chromatin and epigenetics
> Cytoskeletal dynamics
> Developmental biology
> DNA replication and repair
> Mechanisms of human disease
> Mechanobiology
> Membrane traffic and dynamics
> Metabolism
> Nuclear organisation and dynamics
> Organelle biology

Proteolysis and quality control
RNA biology
Signal transduction
Stem cell biology

The journal is clear that the scope of the journal is very broad, and lists just some of its areas of interest.

Some journals might be more specific. Here, you can read the aims and scope of the *Journal of Exposure Science and Environmental Epidemiology*.

The Journal of Exposure Science and Environmental Epidemiology (JESEE) aims to be the premier and authoritative source of information on advances in exposure science for professionals in a wide range of environmental and public health disciplines. The journal is published six times a year, both in print and online.

JESEE publishes original peer-reviewed research presenting significant advances in exposure science and exposure analysis, including development and application of the latest technologies for measuring exposures, and innovative computational approaches for translating novel data streams to characterize and predict exposures. The types of papers published in the research section of JESEE are original research articles, translation studies, and correspondence. Reported results should further understanding of the relationship between environmental exposure and human health, describe evaluated novel exposure science tools, or demonstrate potential of exposure science to enable decisions and actions that promote and protect human health.

JESEE is particularly interested in publishing research that integrates information from exposure science, epidemiology, and toxicology to provide holistic understanding of the most pressing environmental and public health concerns.

JESEE also publishes perspectives, reviews, and analysis of the major advances, trends, and challenges in exposure science for a diverse professional audience. It aims to promote interdisciplinary understanding of exposure science contributions to the environmental and public health field.

The aims and scope here give you a sense of the likely audience, as well as the aims of the journal. The scope is not quite as broad as that of the previous journal, but it still gives clear guidance as to what type of work will be considered, as well as preferred formats.

You can, at this point, send a speculative email to the editor if you choose, outlining your suggested article, and enquiring as to whether it would be appropriate to the journal. This can be a useful first step to take, since the main reason that most journal articles fail at the first hurdle is not due necessarily to a lack of quality, but because they are not a good 'fit' with the journal. In this instance, the submission process is likely to stop with the editor, as opposed to going to peer review. The editor will get back in touch with you with their decision. They might also choose to offer you some general advice going forward as regards a more suitable journal to target, and maybe even some feedback.

At this point, it wouldn't be appropriate to contest the editor's decision on your submission. They are the authority on what would suit their journal, and the chances of your arguments changing their mind are remote.

Work might also be rejected at this stage if there are fundamental flaws or errors in the science of the paper, or if the quality is such that the editor is unwilling to consider it at this point without substantial changes.

If the editor does decide that the article might be suitable for inclusion, then the next step that they will take is to determine the most appropriate peer reviewers, and contact them to see whether they would be willing to review the work.

How does peer review work?

Peer review is the process by which the quality of articles is assessed – an objective evaluation of the value of the work. In peer review, other scientists read your work in order to ascertain its rigorousness, relevance and the nature of its contribution to the field.

Reviewers take several different factors into consideration when making this assessment and thinking about how it might be improved for publication. The following questions for peer reviewers are taken from the Nature Peer Review Guidelines (2016):

- Who will be interested in reading this paper, and why?
- What are the main claims the paper makes and how significant are they?

- Is the paper likely to be one of the five most significant papers published in this discipline this year?
- How does the paper stand out from others in the field?
- Are the paper's claims novel? If not, which published papers compromise novelty?
- Are the claims convincing? If not, what further evidence is needed?
- Are there other experiments or work that could strengthen the paper?
- How much further work would improve it, and how difficult would this be to undertake? Would it take a long time?
- Are the claims appropriately discussed in the context of previous literature?
- If the manuscript is unacceptable, is the study sufficiently promising to encourage the authors to resubmit?
- If the manuscript is unacceptable but promising, what specific work is needed to make it acceptable?
- Is the manuscript clearly written?
- If not, how could it be made more clear or accessible to non-specialists?
- Would readers outside the discipline benefit from a schematic of the main result to accompany publication?
- Could the manuscript be shortened?
- Should the authors be asked to provide supplementary methods or data to accompany the paper online? (Source code for modelling studies, detailed experimental protocols or mathematical derivations.)
- Have the authors done themselves justice without overselling their claims?
- Have they been fair in their treatment of previous literature?
- Have they provided sufficient methodological detail that the experiments could be repeated?
- Is the statistical analysis of the data sound?
- Are there any ethical concerns arising from the use of animal or human subjects?

You can see that this is a lengthy list. However, you can group most of these questions into three main categories that you should bear in mind when thinking about your own work:

- The strength of the paper
- Accessibility/clarity
- Awareness of contribution in relation to field

Types of peer review

While single-blind review is still dominant, the peer review system in the sciences is undergoing some change. We'll give details at the end of the chapter on further reading to help you understand the debate around peer review. For the moment, here are the types of review you might encounter:

Single-blind

Single-blind peer review is the most common type of peer review in the sciences.

In a single-blind peer review, the reviewers are aware of the author's identity, but the author doesn't know who the reviewers are. The rationale behind this approach is that the reviewer is completely free to fully critique the work without any concerns over how the author might respond.

A common criticism of single-blind review is that it does not protect the author from any conscious or unconscious bias that the reviewers might hold, and that the work of a new researcher might be judged differently than someone who is well established and has a prestigious publication history.

The usual response to this is that it's often possible that reviewers are already aware of work that is being carried out in their field, and that they could probably easily identify the author from that knowledge if they chose. They further argue that being aware of the author's previous publications can give them a better sense of their work.

Double-blind

In a double-blind peer review, neither the author nor the reviewers knows the other's identity. This is still the prevailing form of peer review in arts and social sciences. Anonymity is seen as a valuable way of preventing any type of bias influencing the reviewers' decisions, as well as ensuring that the reviewers feel able to fully and honestly critique the work.

As mentioned above, there are those who think that it is likely that reviewers could identify the authors through the topic being studied, etc., and would argue that the double-blind review is not a fully effective method to ensure anonymity.

Open

In an open review, both sides are known to each other – author and reviewers. This type of review is new and still relatively rare. As well as both parties being known to each other, some journals may also publish the reviewers' feedback alongside the finished article.

The rationale behind this is to encourage high-quality peer review by demonstrating its value in how it makes the finished article more robust. By including the reviewers' work in the finished product, it also underlines the importance of peer review, emphasising how the reviewers are part of the piece of work. It also champions the notion of complete transparency throughout the whole process, and allows readers of the journal to see how the article evolved in response to reviewers' comments.

Criticism of this format of review is likely to focus on a lack of protection for reviewers, who may not feel comfortable making highly critical comments on a piece of work that will then be seen by a wider audience. This might possibly lead to a less rigorous review process. There might further be concerns about repercussions for the reviewers who do carry out a critical review.

Pre-registration

Some journals offer pre-registration. Pre-registration is a process where the authors submit aims and methods before data has been collected. If both seem sound, then the journal will agree to publish the end result regardless of results (the work produced must still be rigorous, of course). This practice is currently more common in life sciences, and specifically in subjects such as psychology and neuroscience. The reasoning behind this is that it removes undue pressure from authors to produce positive results, and encourages a focus on transparency and academic integrity.

After peer review

After the article has been peer reviewed, you will receive feedback on your work. This will take the following format:

- Submit as is, no changes required.
- Accepted if the following minor revisions are made.

- Make revisions and resubmit for peer review again.
- Revisions suggested, but the revisions are large enough that resubmission is unlikely in the near future.
- Rejection.

Next steps

You should acknowledge receipt of this feedback and thank the editor for the work of everyone involved in the review process. If you're planning to resubmit the article after making the suggested changes, then you should let the editor know this as soon as possible, and ask about specific deadlines for resubmission.

You might encounter a relatively new practice known as transferable peer review. In this instance, the journal may decide, after peer review, that they do not wish to proceed with publication, but do think that the article might be better suited to another journal. The author is then presented with the option of transferring the work to another journal and going through the peer review process again.

The advantage of transferable peer review is that the author is provided with a speedier second option than submitting their work to their second choice journal: everything is simply transferred to the second journal. It should be noted, however, that there is no obligation on the second journal to publish the transferred work. It simply goes through the usual processes of peer review.

It is most likely, however, that your work has gone through the single-blind review process. You now have your reviewers' feedback. If the work has been accepted without revision, then you have no further work to do. If the work has been given a conditional acceptance subject to some minor changes, then your next step is to make those changes and resubmit. If you have been directed to undertake work that is not feasible at the present time, or given an outright rejection, then you turn to your second-choice journal.

Responding to feedback

If you have been given feedback that indicates publication is possible based on how you respond to the comments offered, then you have more work to do.

When you receive your reviewers' feedback, don't immediately make changes. Read the feedback carefully, and then give yourself some time to absorb them, and think of how you might modify the article based on their comments. Put the article to one side for a few days. It's easy, if you've been particularly invested in a piece of work, to react strongly if you feel that your reviewers' comments were perhaps more critical than you expected, or if you perhaps feel that one or two comments were unfair or a misreading of your work.

It's important, of course, not to respond while in that state of mind – which is why we recommend taking a short breather between reading the feedback and beginning to respond to it. It can also be useful, at this point, to allow someone else to read your work before you've made revisions. It can be easy, especially if you disagree with some of the feedback you've received, to fail to really engage with the suggestions offered, and to only make very superficial changes to the work. However, a failure to properly engage in the peer review process isn't going to strengthen your work, and could lead to the editor declining to continue with the process. As such, it's better, then, to allow someone who is impartial to read your work both before and after you have responded to feedback, and who can let you know whether you have really made significant changes.

Look now at your reviewers' comments. Even if you feel that there are points where they have perhaps misunderstood or misinterpreted what you wanted to say, carefully reread your work and ask yourself why this might be the case. Did you fail to provide sufficient background information to enable them to understand the nature of your contribution? Could there be writing-specific issues that have caused problems? If, for example, you've buried information in long, awkwardly constructed sentences, or in the middle of a dense paragraph, then their attention has perhaps wandered – running the risk that they miss the point you want to make. Was your phrasing somehow ambiguous? Did you try and cover too many points in one paragraph?

Try to ensure that you have an action point for each comment made, and keep track of these. For example:

- **Comment 1** – There's a lack of clarity over how this study builds on previous approaches.

- **Action point 1** – Have expanded section on previous work on page two of the introduction. There are now summaries of key studies on pages 2–4. Further, have reworded concluding section of discussion to clearly state key contributions, and added two paragraphs detailing these contributions on page 11.

- **Comment 2** – The discussion seemed to lack direction, and certain aspects of the data were underexamined.

- **Action point 2** – Have added headings within the discussion to make the flow of discussion more obvious. These headings appear on pages 8 and 9. Have expanded the final paragraph on page 8 to more fully discuss the data obtained in the second trial, and added two paragraphs on page 9 which discuss the full ramifications of the material from the third trial.

As you can see, it's important to take even feedback which might seem vague and try to create specific actions from it.

If there is any point on which you genuinely and strongly disagree with a peer reviewer's comment, and do not wish to make a change that they have suggested, then you can do so – provided that you are able to fully back up your point with appropriate reasoning and explanation. Be aware, however, that the editor has the final say on any decisions, and might well choose to agree with the reviewer, even after you have argued your case. If a compromise cannot be reached, then the work will not be published. This is your call to make, and you are likely to want to talk to other people who were involved in the work before proceeding.

If two peer reviewers disagree on a major point to the extent where they recommend that the writer takes two different courses of action, then the editor might decide to appoint a third reviewer who can take both sides into consideration and offer some sort of compromise solution. Alternatively, the editor might choose to make the final decision themselves.

Checks

The key attributes of a journal article are coherence and concision. You should have one strong contribution/observation/point to make, and this should be reflected in your writing. There should be no digressions, no matter how interesting they might be.

The quality of the abstract is crucial in effectively catching your readers' attention, whether that's the editor who will read it first and decide on whether it's a good fit for the journal, or the readers who will encounter it after publication.

Insights from a journal editor

Whether the abstract is for a journal article, a conference or your PhD thesis, you better get it right because the sad truth is: most people only read the title and sometimes the first few lines of the abstract. As a graduate student I struggled to write abstracts because I had no clue what to include and what to leave out. Later, as an editor, I was often annoyed by the lack of substance and logical inconsistency of many abstracts. Here is my personal recipe for writing abstracts. The main trick is not to start from the beginning; instead, start from the middle.

First, write down what you did. 'We measured/calculated/observed/analysed X.' This is the foundation of the abstract. Then write how you did that. 'We used method Y.' Here you can provide more or less technical details about the methodology depending on the space constraints and type of readership. Next, explain what you found. 'We found out that X is …' Now, go to the very beginning and tell your readers why you bothered to do this research in the first place. Not: 'To get my PhD', 'To have another publication', 'My supervisor said I should do it'; but, for example: 'We don't know the reason for A', 'We can't solve practical problem B', 'We can't explain why C happens'. The final step is to go to the very end and tie things together. The reason why you did the research should be clearly linked to what you found out and its interpretation. For example: 'This gives us a clue for A', 'This provides a solution for B', 'This could help explain C'. Beware not to have a gap between the first and last sentences: your findings should be obviously relevant to the original question. If they are not, then you likely hyped the motivation so it's a good idea to bring expectations down a bit.

In a nutshell, writing an abstract is all about answering these questions in the numbered order and always having a clear link between 4 and 5 even if that makes things a bit less exciting.

4. Why did you do it?
1. What did you do?
2. How did you do it?
3. What did you find out?
5. What does this mean?

Note that the same principle can be applied to structuring an article, oral or poster presentation, or even your thesis.

Iulia Georgescu (Chief Editor, *Nature Reviews Physics*)

One student we worked with uses the word clouds we discussed on p. 141 to help her write her abstract – creating a word cloud of her article draft, looking at the words which appear most frequently, and making sure these words feature prominently in the abstract.

Supporting material throughout should be selected and used judiciously, providing the reader with the understanding and context they need in order to appreciate the contribution that your own work makes. Remember, an article of publication is not a defensive piece of writing like a PhD thesis. You don't have to reassure the reader through extensive writing that you have done all of the requisite reading, nor convince them that you have the skills to critically evaluate others' work.

Sentences conveying take-home messages should be short and impactful, and foregrounded within paragraphs. Paragraphs should deliver one key point, backed up and analysed as appropriate. You have a limited word count, and every sentence and paragraph has to do its job as clearly and concisely as possible.

Read every paragraph in the article. You should be able to write a one-line summary beside it of what you want the reader to take from it. Remember, too, if the first line of the paragraph contains the controlling idea, then these two sentences should be closely related. If not, then it's possible that your paragraph has wandered off the point. This can sometimes happen if you try to cover too much material at once, or if you stray onto a related point. The reverse outline technique discussed in Chapter 7 can be very useful here. If your article is well-structured and written with an eye to concision, then a reverse outline made from topic sentences should read like a bullet-pointed summary of your work.

Even after resubmitting an altered article, the final decision might still be a rejection. If this is the case, then you move on to your second-choice journal. Remember: your time hasn't been wasted. The article that you're submitting now has already gone through peer review, and will very likely be better-written and more robust as a result.

Terminology around publication

This list doesn't intend to fully describe each term. However, we've found that many PhD students are very unfamiliar with much of the terminology around publication. This list gives simple definitions of terms that you are likely to encounter.

> **Impact factor** – The average of how often each article in a journal has been cited in a specific year. The frequency with which articles in a journal are cited is seen as an indication of the importance of that journal in a given field.
>
> **H-index** – A measure of the impact of a specific researcher, as opposed to a journal. Your H-index takes into consideration the number of articles you have produced, as well as how often they have been cited.
>
> **Altmetrics** – These seek to broaden the way in which impact is measured by taking a number of different factors into consideration, such as media coverage, Twitter mentions, citations in policy documents and Mendeley bookmarks. This allows for a different perspective on impact, looking beyond traditional academic contexts to wider concerns. (NB: not to be confused with Altmetric, currently a commercial entity which offers services for a fee.)
>
> **Open Access** – The principle that academic research should be available freely to all. There are different models in place to achieve this aim. Green open access means that the work is freely available through an institutional repository, such as your university's database of PhD theses. Gold open access means that your work is freely available on the publisher's website from the moment of publication.

Other ways of writing beyond the PhD

Academic blogging and Twitter are increasingly popular, and often complementary, ways of engaging with wider audiences. As stated above, altmetrics explicitly recognise the value of such communication by factoring it into a wider picture of dissemination.

As well as a means of publicising published work, blogging is also an informal way to talk to colleagues in the wider community, outside your institution. You can write about progress on your project, about your PhD experience, or about any important issues of the day in your field, or in Higher Education (HE) in general. Blogging like this is another way of making writing an everyday activity, as discussed on p. 20. Further, writing about your research in a more informal, personal tone can also be a useful way of thinking through difficulties.

Twitter is a good way to get you used to talking about your work in this more informal context, and you'll probably notice other PhD students and academics talking about blog posts on Twitter as a way of fostering discussion or, alternatively, discussions on Twitter being used as the starting point for a more in-depth blog discussion. If you're unsure, then start to follow people in your field, or who discuss wider HE topics of interest. You'll soon get a sense of the type of writing that you enjoy, and might want to produce yourself. The hashtag 'phdchat' will let you see what other researchers are talking about, and join in the conversation.

This has the wider advantage of making you feel more integrated in the academic community. One of our students said that she felt much less nervous than she might have expected at her first conference, because she had already talked online with many of the members of the audience. On top of that, you'll find that there are distinct groups within academia talking online, for example, PhD students with disabilities or chronic illness, or students dealing with living in a different country, or students with caring responsibilities.

Experimenting with writing about science in different contexts could also be useful if you are considering a career outside traditional research – maybe in science communication. Even if not, you'll find that expressing your ideas differently and thinking about different audiences will improve your control over your writing, and can help to re-enthuse you about your work when motivation might flag.

CHAPTER 14 **Common Roadblocks During the PhD**

Obtaining a PhD is a complex and demanding process, in which the student undergoes a great deal of developmental change and transition. There's a range of different issues which could cause progress in your studies to stall. In this chapter, we'll look at some of the most common roadblocks you might encounter during your PhD. Some of these issues might be particularly complex, and we'll also recommend extra reading where appropriate.

The PhD is an intellectually demanding apprenticeship in which you are essentially training to become a professional researcher. You are acquiring a great deal of knowledge, critically evaluating the work of your peers, carrying out your research, dealing with unexpected results, and reflecting on the success of your project and its impact on the field.

This apprenticeship involves a change in self-image – making a transition from 'student' to 'professional' – as well as integrating into the academic community, both of your home institution and beyond. This transition can be more complex if you are perhaps coming back into Higher Education after some time away in a professional setting, or if you're returning after having children, or if you're stepping into a different subject area, or if you've had to relocate from another city or country. If you're coming from another academic culture, then you might also have to deal with meeting a different set of expectations.

As well as managing the acquisition of new knowledge and skills and dealing with this transition, PhD students are expected not only to simply 'do the PhD' but are also often encouraged and expected to take advantage of opportunities to both publicise their research project and to develop themselves professionally with their future career in mind: conference attendance, publication, networking, etc.

While this experience is both challenging and enjoyable, it's important to be aware that it also contains potential stress points. Here, we'll consider some of the most common challenges faced by PhD students, and how they can be managed.

- Dealing with your supervisor
- Productivity
- Isolation
- Procrastination
- Perfectionism

Dealing with your supervisor

As we saw in the national guidelines in Chapter 1, successful PhD candidates are expected to demonstrate their ability to work independently, increasingly relying on their own judgement to make autonomous decisions, and to ultimately take responsibility for the project. These skills are developed throughout the course of your studies and, as such, your relationship with your supervisor should evolve to reflect this, with more guidance and advice offered at the outset of the project, and with a more collegiate relationship in place by the end of the PhD.

For many PhD students, especially early in their studies, their supervisor is their connection to the research community and the sole means by which they judge their progress in their work and their development as a researcher (Quality Assurance Agency for Higher Education, Building a Research Community – Student and Staff Views, 14 Aug 2017). If that relationship isn't working the way it should, students can 'stall' in their development, losing confidence in their work and feeling disconnected from the research community in their institution.

We'll now examine some of the challenges you might encounter, and how to remedy them – or avoid them before they can cause problems.

Mismatched expectations

Supervisors and PhD students can sometimes set off on their relationship with very different ideas about their respective roles and responsibilities. While most institutions have guidelines regarding conduct and baseline levels of support, there's still a great deal of variation in, for example, levels of formality within the relationship, the depth of career guidance offered, the nature of feedback offered, and the frequency and duration of supervisory meetings.

If you have the idea that you'll be seeing your supervisor on a weekly basis, and that you'll receive feedback on every submission of written work, detailing writing issues as well as content – while your supervisor has the idea that they will be seeing you on a monthly basis, providing content feedback on relatively polished drafts alone – then it can easily be seen how frustration and a roadblock can develop, with neither side to blame.

It's also important to recognise that differences in expectations can come from different disciplinary backgrounds, different academic cultures and differing understandings of what the PhD process 'should' look like. Even small cultural differences can flag up areas where further discussion might be required. For example:

- My supervisor asked that I call them by their first name. I find it really awkward – you'd never do that at home. You should be respectful in how you address your supervisor.

This might sound like a really small issue, but given that the supervisory model in the UK is about integrating the researcher into the wider academic community, and a relationship which should – by the end of the process – be more collegiate than it is teacher/student, then a student who persists in maintaining a teacher/student relationship may be perceived by their supervisor as someone who is failing to develop in the manner expected of them. If a student is unwilling to challenge their supervisor, to argue a point or to take responsibility for decisions, then from the supervisor's perspective, that would indicate that the development that the UK model expects is not taking place. From the student's perspective, they may well become frustrated and confused about discontent over their progress, as well as uncertain of how to conduct themselves in meetings.

While all supervisory relationships vary, the best way to have a good one is to make sure that you and your supervisor(s) are clear from the outset on the nature of your relationship. This will give you both a sense of what your current expectations are, and enable compromise from the start on issues where you perhaps have very different ideas.

Many universities are aware of how important this initial process is and have, as a result, produced forms which both students and supervisors can complete at the beginning of the PhD. Students and supervisors are usually asked to complete the form by scoring each comment on a Likert scale, which then gives both a sense of where there's agreement, where there's disagreement and where compromise will have to be sought.

Here's an example of some of the questions that might be covered in a form like this.

- It is the supervisor's responsibility to schedule regular meetings.
- The student should send written work at least two weeks ahead of any supervisory meetings.
- It is the supervisor's responsibility to advise students on career planning.
- The supervisor should offer guidance on English.
- The student is responsible for identifying any alternative avenues of support.
- The student should be fully aware of all relevant administrative processes.

If your institution does not have a system like this in place, then you could always draw up your own form and use it as the foundation for an early meeting with your supervisor. If you decide to do this, then the University of Adelaide's 'Expectations in Supervision' form (which can be viewed online) is seen as a good model by many of our colleagues.

Good practice before supervisory appointments

1. **Make sure you send through any necessary documents well in advance**
   If an upcoming meeting is going to be based around feedback on a draft of your work, then it's important to make sure that you've sent this work to your supervisor far enough in advance to allow them time to carefully read it and provide constructive criticism. Supervisors usually have a heavy workload, supervising several students, and they'll find it difficult to give your work the close attention it deserves if you send it to them at the last minute. As a result, submitting a 12,000 word draft at 5pm the day before a supervisory appointment is unlikely to result in a productive meeting.

Try to establish a sense of appropriate time allowances for both of you as early as possible in your relationship. If this hasn't happened, then ask if it would be possible to agree going forward on time frames: how far in advance of any meeting you will send work to your supervisor, and how long your supervisor will need to read written work and provide you with feedback. This will help to avoid frustration on both sides.

2. **Have a clear idea of what topics you'd like to cover/questions you'd like answered**
It's very easy to have an idea of what you'd like to get from your supervisory meeting, but then – once you're actually in the room – for the discussion to move in different directions. It might well still have been a great meeting, with lots of new ideas discussed, but if you don't address important questions and issues, then your progress will be hindered. You don't necessarily need to share this list with your supervisor, but you should write your questions down and make sure you take your list with you. If one of the issues on the list is a 'big' issue, then you might want to let your supervisor know in advance that you'd like to discuss this in the meeting.

Bear in mind that it's not only 'big' issues that can be brought to a supervisory meeting. If anything at all is impeding your progress, even something you think might not be within your supervisor's' interest, it's still worth raising. They might be aware of additional training or other support services within the university that could be of use.

3. **Send a post-meeting email, summarising what was discussed and next steps**
Some universities have established the practice of completing a post-supervision form after every meeting, and which is agreed upon by all present. If this is not in place at your institution, or you would prefer a more detailed record which ensures complete clarity as to what was covered, then perhaps consider an email which follows the following format:

- Thank the supervisors who were present for the meeting.
- Summarise what was covered.
- Detail which next steps were agreed on.
- Set some deadlines.

If not all of the supervisory team was able to attend, then you could consider cc-ing them into the email in order to ensure that everyone is kept fully informed and up to date on progress.

General advice

Be honest about any difficulties you're facing. This includes matters that might seem to you to be outside the PhD. Your supervisor(s) would always prefer to know about anything that might have an impact on your progress, and they're better able to support you in addressing that if they're fully informed as early as possible. Remember – you both want the same thing: a successful PhD.

Sometimes, for a number of reasons, problems do develop whereby student and supervisor(s) cannot reach a solution through compromise. You should make yourself fully aware of all the appropriate procedures within your institution for what to do in these cases. Universities have guidance documents on every aspect of the PhD, and you should make yourself familiar with this documentation. There is also likely to be additional support available to you from student associations within your institution, as well as the Universities and Colleges Union.

Isolation

Studying for a PhD can be an isolating experience. Some researchers first encounter this isolation when they spend time doing fieldwork, or perhaps on a placement at another institution. Sometimes, however, this isolation might not necessarily be the result of working alone or being distant from the institution. Some students might work in a team, but if communication is kept at a fairly superficial level, then it's still possible to feel isolated.

Given that membership in the broader community can help give you a sense of where you are in the PhD process, feeling isolated means that you can't compare your experience with that of other PhD students. This means that you might feel as though you are the only person to encounter problems and issues that are actually fairly commonplace. That can magnify difficulties and make you feel even more isolated.

As such, it's important to try and locate a research community. This is often within your team or within your department, but if that culture isn't in place, then you should seek out alternatives. This might take the form of clubs or societies at your institution. Alternatively, Twitter has a large and diverse academic community constantly discussing not only subject-specific issues, but the larger PhD process and HE in general.

Establishing a writing group can be another good way of forming an informal network around a common goal. These groups don't have to be discipline-specific – just a regular time and place where people come together to write. Everyone benefits from the sense of accountability to the group. On top of this, a writing group allows you to talk to other people about their experience of writing, and of the wider PhD, allowing you to get a better sense of common milestones in the process.

Talking to other people about your work is also a valuable source of feedback. While your supervisor's input is, of course, essential, it's important to have alternative structures in place to provide feedback on your work. If you're working towards a publication, then that feedback might come from peer reviewers. Student-led peer review groups can be a good means of obtaining this kind of feedback, as well as helping you to hone your own skills in review, and can be extended from the type of informal writing group already discussed.

Productivity

There is a huge amount of pressure on researchers to be productive: writing as much as possible, attending conferences, applying for grants, etc. It's a competitive environment, with researchers constantly seeking to distinguish themselves from their colleagues.

However, the PhD is a long and intellectually intense process, and it's natural for your energies and your interest in the work to wane at some points. How can you stay motivated?

Have short-term goals

The PhD is a long haul, and submission after three (or six, if you are a part-time student) years can feel like a long way off, which makes it difficult to feel motivated. Short-term goals, such as presenting a paper at a conference or submitting a piece of work for publication, can help keep you going through patches where your motivation is low. You could also talk to your supervisor about setting yourself 'soft' deadlines: unofficial targets that give you something to work towards.

Acknowledge boredom and tiredness

We've talked to PhD students who feel guilty about being bored with their research. They feel that they've made a serious commitment to their work, and they shouldn't feel bored or tired by it, and that they should feel passionate and energised about their research all the time.

Maintaining a constant level of enthusiasm and engagement is not reasonable or achievable, and nor is working all the time. Acknowledge that there are points where you are likely to feel tired and/or bored and accept that this is a natural part of the process.

Your PhD is a large, complex project, and contemplating it in its entirety can be overwhelming and off-putting. Instead, try to break large tasks down so they feel less overwhelming. For example, if you're writing a chapter that is likely to be around 12,000 words in length, think carefully about how the work might be sub-divided to give you more manageable tasks. Are there, for example, three main sections to this chapter? Immediately, then, you can think about tackling three sections of 4000 words each. Break it down even more if you can.

Use time-blocking techniques to get a sense of your work rate when reading, drafting, editing, etc. This will enable you to set yourself realistic targets. It's easy to set ourselves overly ambitious targets. While it's fine to have high standards in terms of your work ethic, you'll only demoralise yourself if you repeatedly set yourself high targets that you then fail to reach.

Remember, too, that tasks like editing can often take much longer than we think. When we established writing boot camps at our institution, we found that we had participants – staff as well as students – who wanted to use the time for editing, as opposed to producing a first draft. They particularly valued a silent, dedicated time where they could fully concentrate in editing their documents, because they recognised that editing is a demanding and time-consuming task which requires a great deal of effort. Use your thesis journal to get a sense of which tasks are particularly demanding for you, and set aside time accordingly.

Be honest when you set time aside for tasks that have to be carried out. Saying that you'll work 'all weekend' rarely results in working all weekend. You're likely to have other tasks to carry out and multiple distractions that are likely to eat into your time. It's often the case, too, as with contemplating the PhD project as a whole, that facing the prospect of an entire weekend spent writing can actually be off-putting.

Look back at Chapter 2 ('Establishing a Writing Practice') for more detailed examples of how to set yourself up to be productive (or to reset your habits).

Try not to compare your progress to other researchers

Higher Education is a competitive environment. Researchers are judged according to their productivity: how many papers they've published; how much funding they've obtained; how many conferences they've spoken at.

It's important, of course, to be aware of this – especially if you are considering a career in research. However, if – at your current level – you find that comparing yourself to other researchers is detrimental (i.e. if it dents your confidence and paralyses you in terms of progress instead of spurring you on), then it's something you need to address.

Procrastination and perfectionism

These are separate issues, but they can often be closely intertwined – especially in the context of the PhD. Everyone has spells where it feels like work is moving slowly – that they're not as productive as they might like to be. Equally, most researchers – if not all – would profess to having high standards professionally, and to hoping to always produce the best work possible.

This section won't go into a great deal of depth on these subjects. However, there has been a huge amount published on both of them, especially in the context of HE, and there has also been increased focus on them recently as the topic of mental health during the PhD has become predominant. It is important to recognise the role that they can play in your PhD experience, so we'll flag key points here, suggest some strategies to tackle the more straightforward manifestations of these issues and direct you to particularly helpful books, articles and online resources.

Procrastination

In workshops on writing habits, we often ask students what makes them delay writing. There are some very common answers. Avoiding work because you're perhaps bored by a particular topic, or because you don't have good data are frequent responses. Working on a particularly difficult section is another issue. Writing after a supervisor has given negative feedback on a previous piece of work can often prove difficult, or writing when you haven't been given any feedback yet. Students with English as a second language will say that they sometimes put off writing because language issues can make the process difficult and time-consuming, requiring a great deal of concentration. Producing work for a progress review can cause a great deal of anxiety, because your abilities and progress are going to be assessed.

All of this is consistent with the majority of psychologists' views on procrastination: that we procrastinate when there is some kind of negative emotion associated with the task. That might be boredom because it's something you've been working on for a long time; or anxiety, because you're worried about what your supervisor will think of you; frustration, because you don't have good data; and sometimes simply fear – because writing in the past has been difficult and exposed gaps in your understanding, and made you feel like you're not good at what you do.

Understanding that procrastination is a way to avoid these negative feelings and identifying the situations that have produced this negativity offers a way forward.

The first thing to recognise, as discussed in Chapter 2, is that writing *is* sometimes hard. It's often the place where we realise what we still don't know, or where we do our problem-solving, trying to explain exactly how points A and B are linked, or how another theory is relevant to what you want to do. Remind yourself that you're working at the highest level of study possible, and that the work you are doing is demanding. The fact that it feels difficult is not evidence of a lack of ability on your part. It's evidence of the fact that it's difficult. Acknowledge that the work is unlikely to flow easily at first, and that the sticky sections are actually problem-solving activities, not 'writing blocks'.

Procrastinating out of fear of supervisory feedback is understandable. You want to impress your supervisor with your abilities and progress. However, if you try to produce a document with this as your key aim, then the whole process becomes bound up in expectations and self-perception. Remember that your supervisor is trying, first and foremost, to gain an accurate sense of where you are right now. That way, they can give appropriate guidance and feedback that will enable you to develop and move forward. They know that supervision involves guiding a student to become a professional researcher. They aren't expecting you to be the finished product from the outset.

Procrastination needs to be addressed early in cases where language issues are adding another layer of difficulty, because the avoidance of doing any writing practice will only exacerbate those language issues. The best way forward here would be to reflect and develop a clear sense of where the difficulties lie (use of tense, formal language, prepositions), identifying resources to address these difficulties and clearly communicating to your supervisor whether you need additional time to develop your drafts. (For more on this, see the 'Language issues' section, p. 182.)

Having briefly considered some of the causes of procrastination, and some of the ways you might want to begin tackling them, let's think about how they manifest in your working habits, and how you can try to spot them and address them, as well as avoid and minimise them here.

Look out for displacement activities – tasks you undertake that aren't what you set out to do. If we tend to procrastinate when tasks are difficult, boring or anxiety-inducing, then displacement activities tend to be the opposite. They can often be time-limited tasks that are easily achievable and measurable. They're often not as obvious as binge-watching a whole TV series or wasting time online. Instead, they're usually things that make you feel productive, and allow you to convince yourself that you are still working or doing something essential.

For example:

- Answering emails
- Looking for suitable conferences
- Carefully renaming all your files
- Dealing with housework/paying bills
- Indexing your notes

These things will give you a momentary feeling of accomplishment and relief, because you completed something, did something useful, as well as successfully distracting yourself from the task you actually had to carry out. However, once this displacement activity is complete, you're left with even more pressure, because you wasted time you should have spent on what you're supposed to be doing, and now you feel guilty and stressed. This reinforces the negative sensations which we know are associated with procrastination and keeps you stuck in that vicious cycle.

How to tackle it

The types of tasks we procrastinate on are often bigger, more complex tasks which seem intimidating and unachievable. As we discussed, we'll often happily undertake other tasks that we know we can achieve, and that are likely not to take up too much time. This points to one way to move forward...

Break tasks down

Breaking tasks down into small, achievable chunks allows you to make measurable progress and build your confidence. The task will seem less formidable the more you chip away at it. An 8000-word chapter is much easier to work on if you think of it as eight 1000 word blocks. Hitting your target on smaller tasks will motivate you to keep going, whereas a distant goal can be off-putting.

Establish a routine

In Chapter 2, we discussed the importance of reflecting on your weekly routine and figuring out how much time you actually had for writing. With the knowledge of what time is available, set aside and protect specific times as 'writing time'. If you're already gearing yourself up to work on a difficult task, you need to remove as many barriers in your way as possible. Establishing writing as a regular habit in your day will help.

Choose an appropriate space

In Chapter 2, we also talked about choosing an environment which works best for your productivity, and about the distractions that can hinder you. You should try to find a space which minimises possible distraction, and consistently use that space for writing. This underlines the sense of routine mentioned above, and helps you get into the mindset for writing when you are in this space.

Be vigilant about your distractions

We talked above about tasks that seem like work, but are actually only distracting you from the task at hand. Make sure that they don't creep back into your work. If email and social media are major distractions for you, then there are a number of apps available which can either block specific websites at specific times of day, or – alternatively – will completely block you from going online. If you're finding it hard to get a sense of what tends to steal your time in this way, there are also tracker apps that will monitor your time on your computer and then produce a report at the end of the week which tells you how you spent your time. Remember again, while some distractions can still take the form of useful and worthwhile activities, it's up to you to prioritise what actually needs to get done now.

Perfectionism can be closely related to procrastination. Some students will talk about how the need to make the work 'perfect', or the need to have the perfect environment in place, either puts them off starting altogether, or makes for very slow progress. This can sometimes be tied into a reluctance to let their supervisor see work that's any less than 'perfect'.

Comparing your work to published journal articles can often set up unreasonable expectations of how 'perfect' your own work should be. Remember that when you see a published article it has already gone through a number of drafts, redrafts, edits, informal review among colleagues, formalised peer review via the journal and editorial comments. Your own drafts at this point will be very different.

Remember that the PhD is an apprenticeship; a learning process. Nothing is set in stone during the process, and work can always be revised and edited ahead of submission. Some chapters are likely to change a great deal from first draft to final submission.

Since perfectionism often has to do with excessive self-criticism, letting other people see your work can be helpful. Showing your work to your supervisor can feel like a very high-stakes form of feedback, but it's often the only kind of feedback that students receive. The layered feedback approach can be helpful here (Goodson, 2013):

First level

Show your work to someone outside your subject area. They're likely to need you to contextualise your research in clear and simple terms, to define key terminology, and to make links between ideas apparent. Doing this can help you establish a strong foundation, and can sometimes help you to realise what aspects of your work should be foregrounded if you have become lost in the details.

Second level

Show your work to someone in your department – maybe someone you share an office with. They're likely to have a greater level of understanding than the level one reader, but you might find you have to explain some higher-level concepts. They're also more likely to ask trickier questions or point out some reading that you might have missed.

Third level

This is your supervisor, who can offer detailed feedback and content, and offer suggestions about how your work fits into the wider field.

You can see that this approach allows you to gather feedback earlier in the process. You can be reassured knowing that you've improved the work from its first draft stage, and remove some of the intensity involved in having had no one else except you and your supervisor read your work.

Language issues

If English is not your first language, then that can present an additional challenge when studying for a PhD. The key principles that we discussed in other chapters remain relevant here:

Reflection and self-awareness

First of all, it's important to be aware of your own habits, processes and issues. How long does it take you to thoroughly read a journal article?

How long does it take you to produce a draft that can be handed to your supervisor? These are practical considerations that you will have to be fully aware of when managing your time.

Clear communication and shared expectations

As we've discussed above, it's vitally important that any information like this is then communicated to your supervisor, ideally as early as possible. If they know, for example, that you usually prefer to produce three drafts of a document before it is of a standard that you are happy to submit, then they can take this into consideration when setting deadlines.

Planning and implementing a strategy

Some supervisors will be willing to offer detailed feedback on language issues. This can often be useful if you find it difficult to spot problems on your own. However, not all supervisors feel that offering extensive feedback on English issues is part of their job: as we said, expectations of roles within the supervisory relationship can vary greatly. The level of feedback to be expected on this topic should be something that both you and your supervisory team are clear about from the outset of the PhD.

As such, it's important that you make yourself aware as early as possible of *all* sources of training and support within your institution. Support for students who have English as a foreign language varies from institution to institution, as does precisely where it is located: student learning centres, English language units, etc.

As well as identifying sources of support, it's important to recognise self-study that you can undertake. You should be reading as much as possible in your discipline, not purely for content, but also paying attention to tone, vocabulary used, stylistic conventions and structure. Don't just read published journal articles. Try to read other PhD theses, which can give you a much better sense of what's expected of you.

Don't underestimate the power of regular practice and drills. We've worked with students who have achieved significant improvements in their writing by working on weak points with traditional methods such as grammar drills and exercises.

Life outside the PhD

Researchers have lives outside their PhD, and come to the PhD as individuals with their own unique backgrounds and experiences. There is increasing awareness not only of non-traditional PhD students, but also that many students often tackle postgraduate study with additional challenges which can affect their experience, in terms of disability and/or mental health issues.

This is a complex and evolving topic, under regular discussion at the moment across the sector. We can't discuss the topic in detail here, other than to ask that you consider yourself as an entire individual, not just a PhD student, and apply the principles that we've discussed elsewhere in the book: reflect on your experience, identify those areas that are causing difficulties and communicate these as promptly as possible.

It's often the case that the difficulties come not from the researcher, but in the overarching structure of HE not being flexible in terms of accommodating those who do not fit the template of the 'typical researcher'. There are, however, an increasing number of groups seeking to address this via support, and a growing awareness that the idea of a 'typical researcher' and the 'typical experience' should be challenged.

Conclusion

Writing a PhD thesis is a great achievement. It's easy to lose sight of that by the end of the whole process, when you really just want to submit it, but you should take a moment to reflect on what you've actually achieved.

If you're planning to stay in academia, then the skills you'll have developed in writing the PhD, just as in carrying out the rest of the research, are transferable, and will stand you in good stead in your postdoctoral work and career. The writing practice you've established here is something you can carry with you and tailor as your working situation and commitments change.

If you choose to move outside academia, perhaps considering a career in science communication, or in something completely different, then your proficiency in communicating complex concepts clearly and coherently to diverse audiences will still stand you in good stead.

If you're reading this conclusion before you do the bulk of your writing, our final piece of advice would be not to take only our word for it. Make as many connections with other PhD students as you can. Talking to others at your own institution will give you a close network of students going through much of the same admin process as you. Networking with students outside your institution, perhaps through social media, where PhD-related hashtags crop up all the time, will perhaps give you fresh perspectives on what it's like to be a doctoral student as well as highlight the things you all go through in common. Mine those connections for as much advice – direct and indirect – as you can.

Remember: a PhD is the highest academic award that man can bestow upon man.

We wish you well.

Appendix: List of Useful Software and Apps

- **BibTeX** – reference management software that is specifically designed for use with LaTeX.
- **Cold Turkey** – an app which will allow you to block specific websites. You can schedule these blocks, so after looking at RescueTime's reports to get a picture of what distracts you, and when it distracts you, you can block the offending sites.
- **Day One** – a sophisticated journalling app which has several additional features.
- **Diarium** – a popular journalling app which also allows you to dictate entries.
- **EndNote** – widely used reference management software.
- **Journaly** – journalling app which works particularly well across all devices.
- **LaTeX** – a document preparation system which allows for a great deal of control over the eventual look of the document by allowing the user to write the document in markup. Particularly popular with students in STEM subjects, due to the fact that it allows for easy formatting of, for example, mathematical equations, which can often cause problems in Word.
- **Mendeley** – a reference manager and networking tool, which is particularly popular with researchers in the life sciences. Mendeley is owned by Elsevier.
- **MyNoise** – a 'custom background noise machine'. The site offers a variety of sound environments which can be adjusted to suit you. This allows you to both block out distractions and replicate environments which work well for you.
- **MyTomatoes** – one of many websites and apps that work with the Pomodoro™ time-blocking technique. This will time writing blocks, and allow you to save a record of what you achieved in each block. This enables you to track your productivity and get a better sense of, for example, what times and days seem to work best for your writing.
- **Penzu** – a simple journalling site that will send you daily reminders when it's time to write something in your journal.
- **RescueTime** – a time management tool which will track the time you spend on your computer and give you a weekly breakdown of how much

you spent on websites and apps. It will also give you a three-month history, so you can monitor your habits over time. It also allows you to set work goals.
- **WordCloud** – this website allows you to upload a file of any size in order to create a word cloud. You can also generate a list of the most frequently used words in the file.
- **Writefull** – a particularly useful tool for those who don't have English as a first language. Writefull is an extension which will check any phrasing's frequency across Google Books, Scholar, News and Web. The frequency with which the phrasing appears will give you an indication as to whether it's a valid construction in English. It will also show you how the phrase is used in context, and can compare phrases.
- **Zotero** – an open source reference manager.

(All resources mentioned have a free option available.)

Bibliography

Aitchison, C., Catterall, J., Ross, P. and Burgin, S. (2012) '"Tough love and tears": Learning doctoral writing in the sciences', *Higher Education Research and Development*, 31, 4, pp. 435–447.

Bear, M.F., Connors, B.W. and Paradiso, M.A. (2016) *Neuroscience: Exploring the Brain*, 4th edition (Philadelphia, PA: Wolters Kluwer).

Boyle, J. and Ramsay, S. (2017) *Writing for Science Students* (London: Red Globe Press).

Davies, M. (2011) *Study Skills for International Postgraduates* (London: Red Globe Press).

Dunleavy, P. (2003) *Authoring a PhD: How to Plan, Draft, Write and Finish a Doctoral Thesis or Dissertation* (London: Red Globe Press).

Flower, L. and Hayes, J.R. (1981) 'A cognitive process theory of writing', *College Composition and Communication*, 32, 4, pp. 365–387.

Fox, M.A. and Whitesell, J.K. (2004) *Organic Chemistry* (Boston and London: Jones & Bartlett Learning).

Goodson, P. (2013) *Becoming an Academic Writer: 50 Exercises for Paced, Productive, and Powerful Writing* (Thousand Oaks, CA and London: SAGE Publications).

Greetham, B. (2018) *How to Write Better Essays*, 4th edition (London: Red Globe Press).

Heimer, L., Van Hoesen, G.W., Trimble M.D.M. and Zahm, D.S. (2007) *Anatomy of Neuropsychiatry: The New Anatomy of the Basal Forebrain and Its Implications for Neuropsychiatric Illness*, 1st edition (Amsterdam and Boston: Academic Press).

Hubbard, K.E. and Dunbar, S.D. (2017) 'Perceptions of scientific research literature and strategies for reading papers depend on academic career stage', *PloS One*, 12, 12, e0189753.

McCrimmon, J.M. (1984) *Writing with a Purpose*, 8th edition (Boston: Houghton Mifflin).

Mewburn, I. (2013) *How to Tame Your PhD* (London: lulu.com).

Murray, D.M. (1972) 'Teach writing as a process, not product', *The Leaflet*, vol. 74, pp. 11–14.

Murray, R. (2017) *How to Write a Thesis*, 4th edition (London and New York: Open University Press).

Newport, C. (2016) *Deep Work: Rules for Focused Success in a Distracted World* (London: Piatkus).
Pears, R. and Shields, G.J. (2019) *Cite Them Right: The Essential Referencing Guide*, 11th edition (London: Red Globe Press).
Petre, M. (2010) *The Unwritten Rules of PhD Research*, 2nd edition (Maidenhead: Open University Press).
Phillips, E. and Pugh, D.S. (2015) *How to Get a PhD: A Handbook for Students and Their Supervisors*, 6th edition (Maidenhead: Open University Press).
Quality Assurance Agency for Higher Education (2017) Building a Research Community – Student and Staff Views.
Quality Assurance Agency for Higher Education (2015) Characteristics Statement : Doctoral degree from Part A: Setting and Maintaining Academic Standards of the UK Quality Code for Higher Education.
Ramsay, S.W. (2014) 'Studies on an Arabidopsis MYB transcription factor involved in heat and salt tolerance', PhD thesis, University of Glasgow.
Thomas, D. (2016) *The PhD Writing Handbook* (London: Red Globe Press).
Warren, J. (2008) 'How does the brain process music?', *Clinical Medicine*, 8, pp. 32–6.
Watson, C., Kirkcaldie, M. and Paxinos, G. (2010) *The Brain: An Introduction to Functional Neuroanatomy* (Amsterdam and Boston: Academic Press).
Williams, K. and Reid, M. (2011) *Time Management* (London: Red Globe Press).
Wilson, D.A. and Stevenson, R.J. (2003) 'The fundamental role of memory in olfactory perception', *Trends in Neurosciences*, 26, pp. 243–247, https://doi.org/10.1016/S0166-2236(03)00076-6.
Wisker, G. (2007) *The Postgraduate Research Handbook: Succeed with Your MA, MPhil, EdD and PhD*, 2nd edition (London: Red Globe Press).

Index

A
academic databases. *See* databases
Altmetrics, 167, *See also* publication
APA referencing, 63
apps. *See* Appendix, 186–187
articles. *See* journal articles

B
Boolean operators (AND / NOT / OR), 44–47

C
cherry-picking, 82
Cold Turkey, 14
controls, 79–80
critical analysis
   authority of an author, 82
   cherry-picking, 82
   controls, 79–80
   false negatives, 79
   false positives, 80
   optimisations, 80–81
   pressure to publish bad science, 82–83
   recency of cited material, 81–82
   replication / replicability (of experiments), 76–78
   Retraction Watch, 83
   sample size, 76–78
   uncertainty, 84
cross-referencing, 145–146, 151

D
databases, 39–43, *See* Chapter 4
   automated search alerts, 53
   Boolean operators, 44–47
   exclusion criteria, 50
   filtering, 42
   MeSH headings, 53
   PICO, 50
   proximity, 48
   search strategy, 49–50

   sensitivity, 41, 43, 47, 51
   sorting, 42
   specificity, 41, 43, 47, 51
   truncation, 47
   wildcards, 47
digital object identifier, 58
DOI. *See* digital object identifier

E
editing and proofreading, 136–143
   differences between, 137
   effective editing, 137–140
   effective proofreading, 140–142
   general tips, 142–143
   reverse outlining, 138–140
EndNote, 53, 54
exclusion criteria. *See* databases
   confidence, 16
   expectations, 1–9
   independence, 6
   unreasonable. *See* perfectionism and procrastination

F
false negatives, 79
false positives, 80
figures, 147–149
FocusWriter, 15

G
Google Scholar, 39

H
Harvard referencing, 63, 105
H-index, 167, *See also* publication
hypotheses, 32–35, *See also* research question
   hypothesis- led research, 29
   hypothesis-less research, 29
   null hypothesis, 32

I
impact factor, 82–83, 155, 167, *See also* publication

J
journal articles
  finding. *See* Chapter 4: Finding Literature
  navigation and reading sections of, 68–70
  types, 68

L
literature searching. *See* databases

M
MEDLINE, 53
MeSH headings, 53
methods, 95
  vs. methodologies, 36
Mynoise.net, 18

P
paragraphs, 125–35
  common problems, 129–133
  properties, 125–126
  types, 126–128
    argument, 126–127
    contrast/comparison, 128
    detail, 128–129
    process, 127
  using connectives, 133–135
PhD
  by publication, 11–12
  common roadblocks, 169–184
    boredom, 176
    isolation, 174–175
    language issues, 182–83
    perfectionism and procrastination, 177–182
    life outside, 184
    project planning, 29–30, 36
    requirements of, 2
PICO, 50–52
plagiarism, 101–102
  paraphrasing, 102
  quoting, 102
  self-plagiarism, 104–105
  summarising, 102

proximity. *See* databases
PubCrawler, 53
publication, 7–8, 152–166
  factors to consider, 153–154
  motivation, 152–153
  peer review, 158–159
    double-blind, 160
    open, 161
    pre-registration, 161
    responding to reviewers, 161–164
    single-blind, 160
  pressure to publish, 82–83
  process, 155–158
PubMed, 53
punctuation, 115–123
  adjectives, 118–119
  colons, 122–123
  commas, 115–121
    common errors, 119–121
  semi-colons, 121–122

Q
Quality Assurance Agency (QAA), 2

R
referencing
  elements of a reference, 57–59, 100
  formatting a reference list / bibliography, 64
  organising, sorting and annotating sources, 64–65
  reasons, 97–99
  reference management software, 59–66
  styles, 105
    APA, 63
    Harvard, 63, 105
    Vancouver, 63, 105
replication / replicability (of experiments), 76–78
RescueTime, 12, 14, 186
research question, 30–32, 50, 78–79, *See also* hypotheses
Retraction Watch, 83

S
sample size, 76–78
scooping
  being scooped, 52
sensitivity. *See* databases

sentences, 106–115
  how to employ, 111–115
  types of sentence, 106–111
    complex, 109–110
    compound, 107–109
    compound/complex, 111
    simple, 107, 124
software. *See* Appendix, 186–187
specificity. *See* databases
statistical power, 78
study design, 78–79
  controls, 79–80
  optimisations, 80–81
supervisory relationships, 6, 170–174
  good practice for meetings, 172–173
  mismatched expectations, 170–172

T
tables, 147–149
theory
  vs. hypothesis, 35
thesis chapters
  basic principles, 85–86
  body chapters, 91–93
  discussion section, 96–97
  introduction section, 95
  introductory sections, 88–91
  methods section, 95

results section, 96
section numbering, 146
structure, 85–93
typical overall structure, 87–88
truncation. *See* databases

U
uncertainty, 36, 84

V
Vancouver referencing, 63, 105
viva, 96

W
wildcards. *See* databases
WriteMonkey, 15
writing
  environment, 16–18
  establishing a practice, 10–28
  habits
    reflection, 12, 13
  keeping a thesis journal, 20–21
  managing your time, 11–16
    creating time, 13
    dealing with distractions, 14
  motivation, 19
  performing an audit, 11
  process and product model, 21–27

www.ingramcontent.com/pod-product-compliance
Ingram Content Group UK Ltd.
Pitfield, Milton Keynes, MK11 3LW, UK
UKHW021909220326
469204UK00009B/275